Scientific Paper Writing: A Survival Guide

Bodil Holst, Illustrations by Jorge Cham

CreateSpace Independent Publishing Platform

Copyright © 2015 Bodil Holst
All rights reserved. No part of this book may be used or reproduced in any manner whatsoever without written permission.

Published by:
Bodil Holst
University of Bergen
Department of Physics and Technology
Allegaten 55
5007 Bergen, Norway
bodil@cantab.net
CreateSpace Independent Publishing Platform (www.createspace.com)

Cover:
Layout by Christian Bakke, University of Bergen Communication Division.
Drawing *"The Paper Jungle"* by Jorge Cham (www.phdcomics.com).

ISBN-13: 978-1516886265
ISBN-10: 1516886267

Foreword

The objective of this book is to provide you with an entertaining guide on how to write a good scientific paper, but perhaps more importantly, to offer you guidance on how to publish your work in such a way that you get the most acknowledgment for it. Good science will be acknowledged eventually, but the scientific paper publishing system with its journals, editors and peer reviewing is not an infallible black box: It is a system made by humans for humans and as such, it suffers from weaknesses and imperfections. Several great scientific results were initially rejected for publication because the referees failed to see their importance: for example scanning tunneling microscopy (first paper rejected 1981, Nobel Prize 1985) [1] and the giant magnetoresistance effect (first paper rejected 1989, Nobel Prize 2007) [2], just to mention a couple close to my own field. An important aim of this book is to prevent this from happening to you. I want to help you get as much acknowledgment as possible for your work now, when you can enjoy it, rather than after you have died or after your funding has run out.

This book is based on a workshop on scientific paper writing and publishing which I have carried out for several hundred participants at various universities over the last 10 years. The book, like the workshops, is mainly aimed at researchers within the natural sciences and engineering, but I have occasionally had workshop participants from the humanities, social sciences and architecture and they claim that they have benefited as well. I see it as an important justification for my workshops and this book that they are essentially a hobby. In real life I am a professor of physics at the University of Bergen where I run my own research group. I have published in most of the major journals in my field including *Science* and *Nature* (though not as much as I would have liked to). The whole idea of this book and the workshop is to present the paper writing process from the perspective of an active scientist.

Scientific paper writing is a skill. Some people are more talented at it than others, but like any skill it can be learned and practiced. The strange thing is that while it is (mostly) acknowledged by your supervisor that when you embark on your PhD you will need some initial instruction from other people to be able to work with the lab equipment, the computer code, etc., it is often assumed that scientific paper writing is something you can learn on your own "by osmosis" simply by reading scientific papers. This will work eventually and for some better than others, but it is hardly the most efficient way of learning. This book provides you with a tool kit for handling the whole process of writing a scientific paper. It describes the structure of a scientific paper and presents a scheme for drafting a manuscript from the initial idea/question/hypothesis to the final product: a well-structured paper presenting your final results in a clear manner. Interestingly, the final results may have very little to do with the initial ideas! A whole chapter is devoted to the issue of which journal to publish in with special focus on submission to high-ranking journals. The book also contains a section on open access publishing, which is becoming increasingly important. Finally, a chapter is dedicated to handling the referee comments with advice for how an initial rejection can sometimes be turned into acceptance.

During my PhD studies at the Cavendish Laboratory at Cambridge University, I benefited from an excellent PhD supervisor: Bill Allison. I learned the skill of scientific writing in the best possible way: through excellent, patient mentoring. I have observed over the years, however, that many students struggle a lot with scientific paper writing and do not receive the support I was so lucky to get. I also saw that some supervisors were much better at supporting their students than others, thus giving them a much better start in their careers. It seems unfair to me that the choice of supervisor, which many of us make without any regard to this particular factor, should be so crucial for the future career of a scientist. I decided that one aim of my workshop (and eventually of this book) would be to contribute to putting everybody on an equal footing as much as possible.

It is my sincere hope that this book will lower your energy barrier for embarking on paper writing, help you write up your results faster and better and provide you with a useful "behind the scenes" understanding of the whole publishing process. Scientific paper writing is a serious topic, not least because it is so important for a scientific career. Nonetheless, I do not want this book to be too serious. You get an informative, amusing read and I am very grateful to Jorge (www.phdcomics.com) for permission to use his work for illustrating it.

Paper writing is not an exact science. If you follow the tips and guidelines I provide you with here, you will do fine. However, some suggestions are a matter of taste and style and there is not one right way of doing it.

Print-on-demand makes it easy to update a manuscript. I very much appreciate all feedback (positive or negative!) on this book from you, the readers, which will help to improve it. Please send comments to bodil@cantab.net.

Bodil Holst, Bergen, December 2015

Acknowledgments

Firstly I thank my PhD supervisor, Bill Allison, Cavendish Laboratory, University of Cambridge, for providing me with such excellent training and mentoring in science and scientific writing. If all scientists were like you, this book would not be necessary. I thank all the colleagues with whom I have done scientific work and co-authored scientific manuscripts over the years; I can truly say that I have learned from all of you in various ways. Further, I thank all the participants in my workshops for their questions, ideas, feedback, etc., which provided essential inspiration for this book. Special thanks go out to Sean McCarthy, Hyperion Ltd., for being such an excellent role model with his workshops on EU-proposal writing and for giving me initial encouragement and practical advice for my own workshops. I also thank Professor Reitbauer, Technical University of Graz, and Charles Kern, Technical University of Munich, for providing the first testing grounds for the workshops. Thanks go to *Nature* for allowing me to print their guidelines for writing a summary paragraph. I thank the University of Bergen and the Bergen Research Foundation with former Head of Department Professor Jan Petter Hansen, Former University Director Kåre Rommetveit and founder and patron Dr. h. c. mult. Trond Mohn for providing me with a permanent position and start-up grants of a total of 3.5 million euro. Most of my own research referred to in this book could only be carried out thanks to this funding. I also thank the Faculty of Mathematics and Natural Sciences and the Department of Physics and Technology with former Head of Department Professor Geir Anton Johansen for crucial financial support for writing this book through the "Gender Equality Initiatives" (Likestillingstiltak) program. I thank the Communication Division with Christian Bakke for assistance with the cover layout. One of the messages in this book is that a scientific manuscript does not have to be written in perfect English to be accepted. However, when I embarked on this book project, I did feel that writing a book about scientific paper writing is not quite the same as writing a scientific paper. Somehow it seemed to me that a higher standard was required to make the book convincing. I am very happy that I employed Jenny Sandhaas (www.transatlanticlink.com) as editor. I highly recommend her! Finally, I thank Bjørn and Jarl, without whom this book would not have happened now, if ever.

Contents

I. Some Background Information — 1

1. What is a Scientific Paper? — 3
1.1. A Brief Historical Overview — 3
1.2. What is Peer Reviewing? — 6
    1.2.1. Single Blind and Double Blind Reviewing — 7
    1.2.2. A Small Reflection on the Peer Review System — 7

2. Why do We Write Scientific Papers? — 9
2.1. The Noble and the Pragmatic Reasons — 9
2.2. A Small Reflection on PhD Students and Paper Writing — 11
2.3. Quantifying the Quality of a Scientist — 12
2.4. Journal Impact Factor — 13
2.5. Journal Eigenfactor (No, this does not have anything to do with quantum mechanics!) — 15
2.6. Hirsch Index — 15
2.7. Cumulative Citation Index — 16
2.8. A Small Reflection on Impact Factors — 17
    2.8.1. The Upcoming Danger of "Faked" Impact Factors — 18

3. The Journal Jungle — 20
3.1. Different Types of Scientific Papers — 20
3.2. A Small Reflection on Technical Comments and Errata — 22
3.3. What Journal to Publish in? — 23
3.4. Scientific Society Journals Versus Commercial Journals — 25
3.5. Open Access Journals — 26
3.6. Conference Proceedings Versus Regular Journals — 28
    3.6.1. Conference Abstract — 28
    3.6.2. Conference Proceedings Paper — 29

Contents

    3.7. Publishing Interdisciplinary Work 31
    3.8. Maximize the "Findability" of Your Papers 32
        3.8.1. Publication Search Engines/Databases 32
        3.8.2. ResearchGate and Other Online Social Networks 33
    3.9. *Nature* versus *Science* - A PHD Comics Story 34

II. Writing the Paper   38

4. Getting Started   40
    4.1. Why is it so Difficult to Get Started? 40
        4.1.1. A Small Reflection on (Experimental) Scientific Work 44
    4.2. General Tips Before You Start Writing - What is the "Take-Home Message" of Your Paper? 47
    4.3. Pitfalls 49

5. The Structure of a Scientific Paper   52
    5.1. Title 54
        5.1.1. Declarative Versus Neutral Titles (This is Important) 56
        5.1.2. Two Sentence Titles 58
        5.1.3. "Empty" Words in Titles 58
        5.1.4. Using "Dynamic" Rather than "Static" Language 59
        5.1.5. Humorous Titles 59
    5.2. Authors 61
        5.2.1. Who Should be Authors? 61
        5.2.2. In What Order Should the Authors be Mentioned? 63
        5.2.3. Shared First or Last Authorship: A Good Way to Increase Justice 65
        5.2.4. Corresponding Author 66
        5.2.5. Tracing the Work of an Author 67
    5.3. Abstract 69
        5.3.1. Free-Style Abstract 69
        5.3.2. *Nature*'s Summary Paragraph Guideline 71
        5.3.3. Sneezing, Cakes, Coffee: Demonstration Abstract Examples . . 74
    5.4. Key Words and/or Subject Classification Numbers 79
    5.5. Introduction 80
        5.5.1. Motivation for the Work 83
        5.5.2. State-of-the-Art/Previous Work Done in the Field 83

		5.5.3. Paper Outline	86
	5.6.	Methods/Experimental Setup	87
	5.7.	Theoretical Background	93
	5.8.	Results and Analysis	94
	5.9.	Discussion	96
	5.10.	Conclusion	97
	5.11.	Acknowledgments	100
		5.11.1. Author Contribution Statement	101
	5.12.	Citations/References	102
	5.13.	Appendices/Online Supplementary Material	105
		5.13.1. Scientific Fraud - Stopping Schön & Co.	106

6. Drafting the Manuscript: A Possible Approach — 108
- 6.1. Writing the Paper Step by Step — 109
  - 6.1.1. A Small Reflection on the Recycling of Text and Figures — 115
  - 6.1.2. The Guttenberg Case - "Dr." von Copy zu Paste — 116

7. Language: Dos and Don'ts — 117
- 7.1. The Danger of Synonyms — 118
- 7.2. Bigger, Better, Many — 120
  - 7.2.1. Avoid Words with a Specific Meaning Unless you Use Them to Mean Just That — 121
- 7.3. Active or Passive Voice? — 121
- 7.4. Present or Past Tense? — 123
- 7.5. Keep It Short and Simple (Kiss) - A Note Particularly for German Native Speakers — 125

8. A Paper or a Patent or Both? — 126

9. The Random Paper Generator - No this is NOT an Easy Way Out — 128

III. Submitting the Paper — 129

10. Preparatory Steps — 131
- 10.1. Consider the Psychology of the Referee (and the Editor) — 131
- 10.2. Try to Get to Know the Referees and Editors — 134

Contents

11. The Submission Process — **141**

12. The Cover Letter to the Editor — **145**
    12.1. Examples of Cover Letters 146
    12.2. Making it Past the Editor's Wastepaper Basket 149

IV. Publishing the Paper — 151

13. Possible "Return States" of a Submitted Paper — **153**

14. Resubmission or How to Address the Referee Comments? — **157**
    14.1. Some Real Examples of Referee Comments. 158
    14.2. The Reply Letter 159
        14.2.1. An Example of a Set of Real Referee Comments with Reply Letter 161
        14.2.2. How to Handle Some Typical Referee Comments 165
    14.3. Being a Referee Yourself 170

V. Final Remarks — 171

15. A Small, Final Reflection — **173**

Bibliography — **174**

Index — **177**

Part I.
Some Background Information

Chapter 1.
What is a Scientific Paper?

1.1. A Brief Historical Overview

It is interesting to take a brief look at the history of scientific publishing, because it gives some insight into why scientific publishing is done the way it is. This is only a brief, non-scholarly introduction to a very large and complex topic and it can easily be skipped if you are not so interested.

Until around 1450, the only written documents available were hand written. A main purpose of lectures at universities in the early medieval times was to allow the students to copy the professor's handwritten manuscripts, which they would not otherwise have had access to. Around 1450 Gutenberg invented the printing press which suddenly made books available at a much cheaper price and in much larger numbers [3]. It is amusing to note that almost 600 years later many professors have still not realized this change of affairs. They continue to give their lectures in the early medieval fashion simply writing the content of their lecture notes on the blackboard (or even worse, projecting them in PowerPoint onto a screen!) for the poor students to copy.

Scientific breakthroughs in the Renaissance were communicated through printed books. Despite the invention of the printing press, it was still expensive to publish and often, if the authors were not wealthy themselves, they had to rely on wealthy patrons to promote their work and pay for publishing.

The origin of scientific papers/articles in the form we know today can be dated back to 1665 when *Philosophical Transactions* was published for the first time as the first journal in the world exclusively devoted to science. It has been published continuously ever since, making it not only the oldest but also the longest running scientific journal. *Philosophical Transactions* is published by *The President, Council, and Fellows of the Royal Society of London for Improving Natural Knowledge* (short *Royal Society*). The *Royal Society* is the oldest scientific society still in existence [4]. It dates its origin back to 28 November 1660 when a group of 12 men gathered after a

Chapter 1. What is a Scientific Paper?

lecture by Christopher Wren and decided to meet weekly to witness experiments and discuss scientific topics. The motto of the *Royal Society* is "*Nullius in verba*" which roughly translates "*take nobody's word for it*". Many prominent British scientists have been *Royal Society* Fellows and many prominent scientists from the rest of the world have been associated with the *Royal Society* as foreign members. One of its first presidents was Isaac Newton [5].

The first person in charge of organizing the weekly experimental sessions at the *Royal Society* was Robert Hooke. Hooke was a great scientist of many achievements. He is famous for the law of elasticity which bears his name [6]. He was also the first to use the word "cell" for biological cells when describing a piece of cork-wood that he observed under a self-built microscope. The cell structure of cork reminded him of monk's cells, hence the name [7].

All articles ever published in the *Philosophical Transactions* are available online. In 1886 the journal was divided into two journals which were very imaginatively named *Philosophical Transactions A* (covering the physical sciences) and *Philosophical Transactions B* (covering the life sciences).

The introduction to the very first volume of *Philosophical Transactions* was written by the first editor, Henry Oldenburg [8]. Despite the rather convoluted use of the English language (to our ears), the content is so beautiful. Almost 350 years later, it still so perfectly describes the most noble aims of scientific work and publication, that I reproduce it here in its entirety (I promise, this will be the only citation of this kind, so please do not let it scare you off!):

> *Whereas there is nothing more necessary for promoting the improvement of Philosophical Matters, than the communicating to such, as apply their Studies and Endeavours that way, such things as are discovered or put in practise by others; it is therefore thought fit to employ the Press, as the most proper way to gratifie those, whose engagement in such Studies, and delight in the advancement of Learning and profitable Discoveries, doth entitle them to the knowledge of what this Kingdom, or other parts of the World, do, from time to time, afford, as well of the progress of the Studies, Labours, and attempts of the Curious and learned in things of this kind, as of their compleat Discoveries and performances: To the end, that such Productions being clearly and truly communicated, desires after solid and usefull knowledge may be further entertained, ingenious Endeavours and Undertakings cherished, and those, addicted to and conversant in such matters, may be invited and encouraged to search, try, and find out new things, impart their knowledge to one another, and contribute what they*

can to the Grand design of improving Natural knowledge, and perfecting all Philosophical Arts, and Sciences. All for the Glory of God, the Honour and Advantage of these Kingdoms, and the Universal Good of Mankind.

"..and those addicted .. may be invited to search, try and find out new things, impart their knowledge to one another, and contribute what they can to the Grand design of improving natural knowledge .." .The essential message of Oldenburg's beautiful introduction is that the new journal is a vehicle for all people to share the knowledge they have gained, even though they cannot attend the weekly demonstrations in London. Note the emphasis on internationalism. The pursuit of scientific knowledge is a concern of *"this Kingdom, or other parts of the World"* for *"the Universal Good of Mankind"*.

The Royal Society was quickly followed by other societies of a similar nature throughout the rest of Europe. *Academie Francaise* was founded in 1666, *Deutsche Akademie der Naturforscher Leopoldina* in 1687, etc.. One may say that the foundation of scientific societies became a fashion in Europe in the 17th and 18th centuries and most of them published their own journal(s).

Some time after the scientific societies started to publish scientific journals, the first commercial publishing houses followed. To this day, commercial journals and journals published by scientific societies exist side by side. *Annalen der Physik,* in which Einstein published his famous 1905 papers, is an example of an early, commercial journal. It was published for the first time in 1799. The famous, world-leading journal *Nature* is purely commercial and was published for the first time in 1869. *Elsevier* publishing company, which presently publishes more than 1800 journals and 2200 books a year, was founded in 1880.

Scientific journals make it easier for authors to publish their work - a major advantage over books. Many early contributions to *Philosophical Transactions* were made by authors of rather humble origin, who would not easily have been able to finance the publication of a book.

Even after the establishment of scientific societies, scientific journals were not the only format for scientific publishing. Many important discoveries and inventions were only published as patents or kept secret and not published at all. This is still true in modern times. Books have also remained popular. Universities started to publish books very early and have continued to do so. Books were published through private means and scientific societies sometimes published books written by their members in addition to their journals. For example, Newton's famous work *Principia Mathematica* was published as a book by the Royal Society in 1687 [9].

Chapter 1. What is a Scientific Paper?

In fact it was only in the late 19th/early 20th century that scientific journals became the standard way to publish new results in the natural sciences. To this very day, it is common in the humanities and social sciences to publish new results in books rather than papers. Of course books are still important in the natural sciences. More scientific books are published now than ever before, but as a general rule they mainly communicate results that have previously been published in scientific papers.

Another thing that has changed over the years is the language used to communicate scientific results. For several hundred years the chosen language was Latin. The journals of the scientific societies were important also in that they allowed people to write in other languages, thereby enabling people who had not received an (expensive) classical education to participate in the scientific discourse. One of the last important scientific results to be published in Latin was the discovery of electromagnetism in 1820 by my fellow countryman H. C. Ørsted. He published the results himself in a short Latin script, which he distributed to scientific societies and prominent scientists of the day [10].

Broadly speaking one may say that since the Second World War, English has established itself as the international, scientific language for the natural sciences and to some extent also for the humanities and social sciences. My sincere advice is: **You better publish your results in English if you want full credit for them**. There are several examples of scientists who did not receive the credit they deserved because their results were published in other languages. This was, for example, an issue during the Cold War, when scientists from Eastern Europe had to publish their results in journals written in Russian. Admittedly, there are also examples of scientists who did not receive the credit they deserved even though they published in English. Hopefully, this book can help you avoid that situation.

1.2. What is Peer Reviewing?

The publication in scientific journals is closely associated with the term "peer reviewing". Peer reviewing refers to the process used to assess the quality of a scientific manuscript that has been submitted for publishing in a journal. The idea is that before a scientific manuscript can be published as a paper in a journal, the quality of the manuscript should be assessed by one or more independent experts in the field - the "peers".

The word "peer" originally means a member of the British nobility, the "peerage". The holder of a peerage is called a peer. Later a peer group came to refer to a group of people who are in some sense equal. When the word peer is used in the context of

1.2. What is Peer Reviewing?

reviewing, it means a person who is (or should be) your equal in terms of education and scientific background and therefore capable of judging the quality of your work.

Peer reviewing is as old as scientific journals. It was introduced by Henry Oldenburg, the first editor of the first scientific journal, *Philosophical Transactions*, discussed in the previous section. Traditionally referees do not receive any payment for their work. Acting as a referee is considered a service a scientist owes the scientific community.

Nowadays a paper manuscript will typically be reviewed by 1-3 referees selected by the editorial staff or the editorial board of the journal, depending on how the journal is organized. The editorial staff will usually be comprised of people who have a scientific training but who are working full time for the journal and are no longer active scientists. The editorial board on the other hand will typically be people who are active scientists.

Some journals send nearly all manuscripts for external peer reviewing. Others, typically high-ranking and hence very popular journals, have an initial screening process whereby perhaps only 10% of the submitted manuscripts are sent for peer reviewing. The initial screening is usually done by the editorial staff.

1.2.1. Single Blind and Double Blind Reviewing

Modern peer reviewing in the natural sciences is usually performed as "single blind" reviewing. This means that the referees know who the authors are, but the authors do not know who the referees are. In "double blind" reviewing, on the other hand, both authors and referees remain anonymous to each other. Double blind reviewing is used by some journals in the humanities and social sciences.

1.2.2. A Small Reflection on the Peer Review System

A lot has been said about the peer review system and it is clear that it is not perfect. It has been argued that the system favors people who are already famous scientists and/or scientists from famous scientific institutions and is also accused of opening the door for nepotism and subjective and erroneous judgments. Further it has been argued that it is very difficult to ensure that referees are really qualified to do their work and that there is not really any incentive for them to do it well other than their scientific conscience. After all, the work is anonymous and usually not paid. Although it is a good thing to be able to list on your CV that you are a referee for a number of journals, it is hardly the most important point. As an attempt to make good referee work more rewarding, the *American Physical Society* has started to hand out official

Chapter 1. What is a Scientific Paper?

"outstanding referee" awards to selected referees doing a particularly good job, the idea being that this is a prestigious award which enhances your CV.

You will hardly find any established scientist who cannot list several bad experiences with referees. On the other hand, many of us have also had very good experience. Along with a couple of bad referee experiences myself, I can honestly say that I have also had papers that have benefited so much from the referee comments that I should have liked to include them as co-authors. Personally, I think that the peer review system, while by no means perfect, still seems the "least worst" system. I do wish, however, that double blind refereeing would be introduced for all journals. In my opinion this would decrease the "established scientist/established scientific institution" effect. People have argued that it is usually possible to guess who has written an article anyway thus eliminating the need for double blind refereeing. This seems a very weak counter argument to me!

The main thing to bear in mind as a scientific paper author is that the peer review system is not perfect. The referees are human beings and human beings are per definition not perfect and influenced by many factors. This means that **the referee comments you receive are NOT the infallible, objective truth**. In fact they may be quite erroneous. In the foreword I mention a couple of famous Nobel Prize discoveries whereby the initial papers were rejected by the referees. The "human factors" in peer reviewing are discussed further in Part IV.

Chapter 2.
Why do We Write Scientific Papers?

2.1. The Noble and the Pragmatic Reasons

Why do we write scientific papers? We may split the answer to that question into two parts:

- The noble reasons:
    - We write scientific papers to communicate our scientific results to the world for the benefit of humanity. This may seem a bit high strung, but for many scientists, including me, this is a strong driving force. I am proud to be a scientist and proud that I am contributing to bringing humankind forward. I honestly believe that the scientific papers I have co-authored are my gift to humanity. Some of the results I have participated in generating I find so beautiful that they fill me with a particular joy and a feeling of having given something special to the world. Writing a scientific paper is a unique chance at immortality. As long as the civilized world exists, your results will be out there in libraries and electronic databases, available to the rest of humankind together with the works of Einstein, Bohr, Curie, and all the others.
    - It is very important to remember that if you do not publish your results, they cannot benefit anybody (except perhaps close colleagues) and are not accessible to the rest of the scientific community. To quote George M. Whitesides, one of the most prolific and highly-cited scientists of our times: *"Papers are a central part of research. If your research does not generate papers, it might just as well not have been done. "Interesting and unpublished" is equivalent to non-existent"* [11].

Chapter 2. Why do We Write Scientific Papers?

- The pragmatic reasons:
  - We write scientific papers to show that we are active, prolific scientists. Publications in peer-reviewed journals remain the main channel for demonstrating scientific excellence. Patents have become increasingly important in recent years, but the acceptance of a patent cannot be compared to the publication of a paper. The patent authorities do not check the quality of your work the way a referee for a journal is supposed to do. The patent authorities mainly check that there is not already a patent covering the invention you propose. In fact, it is perfectly possible to patent some untested idea which eventually turns out to be completely impossible to realize. Large companies sometimes play deliberately on this to confuse the competition.
  - Publications are particularly important at the early stages of a scientific career, because a good publication list is necessary for securing a permanent position. Of course publications are also important later on. Most crucially: if you do not document through scientific publications that you are a high-quality scientist, you will not receive funding for new research and eventually you will get into trouble with your university. Without publications you will also not receive any nice invitations for talks at seminars and conferences, etc.. **Your scientific prestige is defined by your papers**.

2.2. A Small Reflection on PhD Students and Paper Writing

It used to be the case in most countries that PhD students did not need to worry about publications. It was the job of the supervisor to write up the results and it was not a major problem if no publications resulted by the end of the PhD work. Nowadays the situation is quite different. In many universities, in order to get a PhD a student should have at least one first-author paper accepted for publication in a peer-reviewed journal at the time of submission of the PhD thesis. In some cases the PhD thesis can be what is now often referred to as a cumulative PhD thesis. Instead of writing a so-called monograph thesis, documenting your work step by step in a single, coherent manuscript, the cumulative PhD thesis consists of an introductory text followed by 4-5 papers published in or accepted by peer-reviewed journals. Normally the PhD candidate should be the first author of these peer-reviewed papers (see Section 5.2.1).

Scientific paper writing is such an important part of science that it is clearly sensible to be introduced to the topic at an early stage. This would seem to speak for a cumulative thesis, but it should be considered that not all thesis topics are suited for a cumulative PhD thesis. For example, projects that involve setting up a completely new experiment will leave the candidate with much experience and deep understanding even though they often lead to only one paper at the most. On the other hand, the lucky candidate that enters a group at the right time when the experiment is just up and running will be able to produce 4 papers relatively easily. Fortunately, the trend of recent years towards having everybody write a cumulative PhD thesis seems to have abated to some extent. If cumulative PhD theses were made obligatory, it would limit the range of possible PhD topics considerably. As one of my colleagues put it, perhaps a bit harshly: *"If the students have to do a cumulative PhD thesis, it is not possible to let them do any real science"*.

It remains as a general rule, however, that if you are a PhD student and you want to do a postdoc after your PhD: the more impressive your publication list, the better your chances of getting a good position. Therefore, if you have worked on a very difficult and ambitious task which has not led to many publications, it is a good idea to state this openly in your postdoc application and explain why you are damn good even if you have written just one paper.

Chapter 2. Why do We Write Scientific Papers?

2.3. Quantifying the Quality of a Scientist

As mentioned in Section 2.1, publishing in peer-reviewed journals is the main channel for evaluating the quality of a scientist. So how is the quality of a scientist's publication list actually judged? The two main criteria are: number of papers and quality of papers.

- Number of Papers:
  - You sometimes hear people saying: "*It does not matter how many publications you have as long as they are of high quality*" – Forget it! If people look at your publication list and you have 10 publications they will be more impressed than if you have 5 publications (unless those 5 publications are all in *Nature* or somewhere similar). That said, it is true that there is a general move away from quantity towards quality. For example, some funding bodies only allow you to list say the 5 most important publications from your project as outcome in the final project report. This principle is now also applied for some grant applications where you are only allowed to list say your 10 most important publications in your CV. This increasing emphasis on quality papers is very much to be welcomed. It should not create the impression, however, that only really ground-breaking results are worthy of publishing. The problem in the scientific community is not

that too many papers are being published but that too many badly written papers are being published.

* It is always good to publish a new result in a well-written paper even if the result may not seem earth-shattering to you. Often these minor results can help other scientists to progress in their daily work. If your paper can save just one other scientist some time and effort, then it has been worth publishing and remember – you cannot always predict yourself for whom your result may be useful.

- Quality of Papers:
    - The quality of a paper is judged by:
        * The impact factor of the journal where the paper is published. The impact factor of a journal is a measure of how often papers in the journal are cited (see Section 2.4).
        * How many times the paper is cited by other papers. This is quantified using the Hirsch index and the cumulative citation index (see Sections 2.6 and 2.7).

2.4. Journal Impact Factor

The impact factor of a journal is a measure of how often the papers in that journal are cited. The impact factor was introduced by Eugine Garfield in 1955 in an article in *Science* [12]. The first *Journal Citation Reports* was published by the commercial, privately owned *Institute for Scientific Information* (ISI) in 1961 (founded by Eugine Garfield). In 1992 ISI was purchased by the Thomson Reuters Corporation who runs the *Web of Knowledge* with the *Web of Science*, arguably the most important search engine/database for general scientific paper search (see Section 3.8.1). If you have access to the *Web of Knowledge*, you automatically have access to the *Journal Citation Reports*.

When "The Impact Factor" is referred to without any further explanation, it is usually the *Journal Citation Reports* from the *Web of Knowledge* which is meant. *Journal Citation Reports* only publishes impact factors for journals and conference proceedings that are included in the *Web of Science* and only journals and conference proceedings included in the *Web of Science* are used to calculate the impact factor. See Section 3.6 for a discussion of the difference between conference proceedings and regular journals.

Chapter 2. Why do We Write Scientific Papers?

- The established way of calculating the impact factor is quite simple. As an example we take the impact factor for *Journal of Microscopy* for 2006: 1.95. This number was calculated as follows:
  - The total number of papers published in the *Journal of Microscopy* in 2004 and 2005 was 143+101=244
  - The total number of citations in 2006 of *Journal of Microscopy* papers published in 2004 and 2005 was 310+165=475. That is to say, 475 papers published in acknowledged scientific journals in 2006, including the *Journal of Microscopy* itself, cited *Journal of Microscopy* papers from 2004 and 2005.
  - The impact factor for the *Journal of Microscopy* for 2006 is the ratio between the two numbers: citations to recent articles/numbers of recent articles = 475/244 =1.95

We see that the impact factor is actually a very volatile number, subject to quick changes. We also see that the impact factor depends on which scientific journals are considered "acknowledged". The "original" impact factor is calculated on the basis of journals that are available in the *Web of Science*. According to the official website of *Thomson Reuters* on the journal selection process for the *Web of Science*, the *Web of Science* presently includes more than 12000 journals from all areas of the natural sciences, engineering, social sciences, arts and humanities. Every year about 2000 journals are reviewed for possible inclusion in the *Web of Science*. About 10-12 % are eventually accepted. Journals already included in the *Web of Science* are reviewed at regular intervals and occasionally journals not living up to standards are excluded, though Thomson Reuters does not provide any information as to how common that is. See [13] for a discussion on how journals are selected for acceptance in the *Web of Science*.

The impact factor includes self-citations. This means that if a paper in the *Journal of Microscopy* published in 2007 cites two *Journal of Microscopy* papers from 2005-2006, this paper will at the very least be "impact factor neutral". The journal can publish the paper without any risk of lowering its impact factor. It should be noted however that Thomson Reuters explicitly states that exaggerated self-citation is a reason for excluding a journal from the *Web of Science*.

- In 2010 85% of the journals in the *Web of Science* had a self-citation rate of less than 15%: "*Significant deviation from this normal rate, however, prompts an examination by Thomson Reuters to determine if excessive self-citations are being used to artificially inflate the impact factor. If we determine that self-citations are being used improperly, the journal's impact factor will be suppressed for at least two years and the journal may be considered for deselection from Web of Science."* (quoted from [13]).

Nature had the highest impact factor of all journals in 2014: 42.351 according to the journal website. It is difficult to say in general what a good impact factor is; the highest impact factors for physics journals are around 20 (*Nature Physics*, 2014) and 7.5 (*Physical Review Letters*, 2014). In the life sciences, where the research communities are generally larger than in other disciplines, several journals have very high impact factors, for example in 2014 *Cell* had an impact factor of 32 and *Lancet* an impact factor of 39. In Chemistry, journals like *Journal of the American Chemical Society* (JACS) and *Angewandte Chemie - International Edition* had impact factors in 2014 of around 12 and 11, respectively. In some areas where there is not a long tradition of publishing in scientific journals, such as many areas within engineering, you will be hard pushed to find any journals with impact factors much above 1.

2.5. Journal Eigenfactor (No, this does not have anything to do with quantum mechanics!)

The journal impact factor described in the previous section is the most widely used method for evaluating the importance of a journal, but it is by no means the only one. Several alternatives have been proposed. I'd like to present just one here: "The Journal Eigenfactor", which also has the advantage that it is accessible for free (www.eigenfactor.org). The Journal Eigenfactor was introduced in 2007 by Jevin D. West, Carl Bergstrom and others. It is sponsored by the Bergstrom Laboratory at the Department of Biology at the University of Washington [14].

The Journal Eigenfactor also uses citations as the basic measure of impact. However, not all citations count equally. A citation in an influential journal which is itself cited a lot counts more than a citation in a less-cited journal. Furthermore, if a paper only cites a few papers, these cited papers are more important than those in a paper with many citations.

2.6. Hirsch Index

The so-called Hirsch index (h-index) or Hirsch number was introduced in 2005 by Jorge E. Hirsch [15] as a way to measure your productivity as a scientist: Did you

Chapter 2. Why do We Write Scientific Papers?

have just one high-flying paper with many citations or do you consistently write papers that get lots of citations. The Hirsch index, h, is calculated as follows:

- A scientist has Hirsch index h if h of his/hers N papers have at least h citations each, and the other (N - h) papers have at most h citations each. It does not matter if you were single author, first author or co-author on the papers.

    - This means that if you for example have 5 papers with 5 citations or more, you have a Hirsch index of 5. If you have 4 papers with 5 citations, 1 paper with 100 citations and the rest of your papers have 4 citations or less, you also have a Hirsch index of 5.

    - Various online tools allow you to calculate your Hirsch index (*Web of Science, Google Scholar,* etc.). They will not necessarily give you the same number; *Google Scholar* normally includes more citation sources than the *Web of Science* (see Section 3.8.1).

    - No doubt it is just a matter of time before somebody introduces the modified Hirsch index, which would exclude self-citations (citations where a scientist is co-author on a paper that cites his or her previous work), but until that happens, do not forget to cite yourself when appropriate.

    - Get your friends to cite you.

The rule of thumb is that you should try to have a Hirsch index which is at least equal to the number of years you have spent in science, counting from the day you started your PhD. Do not worry too much about this, but do not forget to cite yourself! Theoretically, if you produce two papers a year citing all your previous papers, you can get a Hirsch index equal to the number of years spent in science without anybody else taking an interest in your work.

2.7. Cumulative Citation Index

The cumulative citation index is simply the total sum of all citations of all a scientist's papers in acknowledged journals. Unless explicitly mentioned, the cumulative citation index includes self-citations. Sometimes only the last 5 years are used as a basis for the calculations to assess if the scientist has remained productive, though citations of older papers are also included.

2.8. A Small Reflection on Impact Factors

A lot of criticism has been raised as to the judging of scientists on the basis of impact factors. For example, people argue that this system favors work in "safe fields" with many participants, because this is likely to produce many publications and many citations rather than promoting pioneering, high-risk work. Speaking from my own experience: One of my main fields of research is the development of a new type of microscope using neutral helium atoms as the imaging probe. This is not the kind of work that leads to many publications and the papers we do publish are related to instrumental development and hence not likely to attract many citations. It is also difficult to "sell" a great breakthrough in the development to a high-ranking journal, because the scientific impact will not come until the instrument has matured and can produce images unable to be obtained with other techniques. Thus, basically, if the impact factor was all that counted it would not be good for me.

It is also often argued that it is not necessarily the scientists best at writing papers and "selling" their results and ideas who actually do the best science and have the deepest insight.

This problem is reinforced by the fact that in many situations, scientists are judged by other scientists not exactly in the same field (often happens with grant applications, for example). Hence these other scientists cannot really judge how difficult it might be to obtain a specific result. Rather than scrutinizing the publications in detail, they will tend to look at numbers such as the Hirsch index, journal impact factors, the cumulative citation index, etc., because these can easily be compared. Whether this comparison is just and reasonable is another matter. Calling in absolute experts does not necessarily solve the problem. It is by no means so that scientists with good knowledge in a field always provide a fair judgment. Scientists are only human beings and unfortunately, there is a lot of power struggle and irrationality also in the scientific community.

The increasing emphasis on numbers such as impact factors and the Hirsch index clearly has a negative effect on science by making it very difficult for more eccentric characters to find a place in academia. When I studied in Cambridge, one of the most beloved and respected lecturers at Trinity College was Jeremy Maule, who had not even written a PhD thesis. Trinity College, Cambridge, is one of the most famous academic institutions, counting Isaac Newton, James Clerk Maxwell and J. J. Thomson among its former members. It is also one of the richest academic institutions in the world, and hence it can indulge in employing a lecturer under circumstances which in this day and age would usually not be possible. Jeremy Maule was a humanities

Chapter 2. Why do We Write Scientific Papers?

scholar. He had read literally everything and spent a large part of his time listening to students and other scholars asking him for advice, which he gave freely and generously. He himself referred to these discussions as "high gossip". He was a source of immense inspiration to many people and his death was a deep loss to the research community. For a moving obituary see [16]. The possibilities for people like Jeremy Maule to find their way in academia are becoming increasingly limited and this is clearly a great loss. On the other hand, the increasing importance of quality based on well-defined, measurable criteria should, in principle at least, enforce fair competition and limit nepotism, which is clearly a good thing.

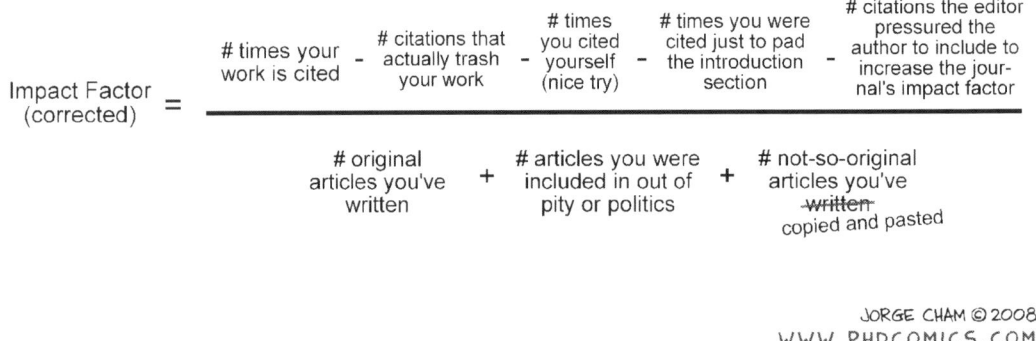

2.8.1. The Upcoming Danger of "Faked" Impact Factors

All the arguments above aside, there is an additional danger connected to the concept of the impact factor as it is being administrated at the moment. The word "university" is protected in most countries. It is not possible to set up a university without approval from official authorities. This does not apply to words such as "research institute", "science foundation" or "think tank". Anybody can use these words and anybody can start a journal with a scientific sounding name. A so-called "open access" journal (see Section 3.5) requires only a website and some server storage space. So it is clear how it would be very easy and cheap to start up a set of journals which nominally follows all the rules of scientific peer reviewing and at the same time deliberately pushes each others nominal influence by deliberately citing each other. This would not be directly

2.8. A Small Reflection on Impact Factors

spotted by the *Web of Science*'s self-citation check (see Section 2.4). Such journals could be used to push the careers of individuals and/or the publication of what one might label "biased" results.

The danger of biased journals is not new. In the early years of the debate on the danger of tobacco, the tobacco industry supported several so-called "research institutes". Journals associated with these institutes published nominally scientific papers questioning whether the danger of tobacco was really genuine. It appears that the same principles are being applied to the climate debate. For a very interesting discussion of this see [17]. The problem is potentially even more serious nowadays because it is so easy to start an open access journal.

Chapter 3.
The Journal Jungle

3.1. Different Types of Scientific Papers

The basic idea behind all scientific papers (except review papers) remains what it has always been: to communicate a discovery to the scientific community and enable interested readers to reproduce the results themselves. Despite the fact that the basic idea behind all papers remains the same, different types of results may be presented with different emphases and thus, over the years, different types of papers have emerged. The different paper types can be separated into a range of groups as listed below. These are just guidelines rather than formal distinctions and the exact names may differ from journal to journal and discipline to discipline. Some journals specialize in one type of paper, others have several different types. You should not consider any of the definitions and length estimates below as "strict"; rather they are intended to inspire you when you are thinking about how to organize the presentation of your results in one or more papers.

- *Letter/Rapid Communication/Short Communication*: This is a short paper (typically up to four journal pages) communicating a single, important result. Some journals publish other types of papers as well as letter papers, but there are also several journals specializing exclusively in letter papers. These journals tend to be quite prestigious and have high impact factors. Examples of journals specializing in letter papers are: *Physical Review Letters, Applied Physics Letters, Nanoletters, Chemical Physics Letters*, etc..

- *Full Papers*: These are longer papers (typically up to 20 journal pages) communicating a single scientific topic in more detail. There will usually be one or maybe two main results and several minor results. Often, but by no means always, a full paper can be based on several letter-type papers. A good way of organizing your results can be to write the main results in individual letter

3.1. Different Types of Scientific Papers

papers first and then compose a detailed overview with some additional new results in a full paper. A full paper must always contain some new results in the form of new measurements or more detailed calculations, etc.. *The American Physical Society* has as its flagship journal *Physical Review Letters*, containing short communications (4 pages max) on important results of general interest to the physics community. In addition this Society publishes several other journals with typically longer papers on more specialized topics (*Physical Review A, Physical Review B*, etc.). The idea is that the most important results go into *Physical Review Letters* in a short, concise form, and the other journals are open for more extensive descriptions. Sometimes you do find shorter papers of 4 pages and less in the other journals as well. When that is the case, these are often papers which were initially submitted to *Physical Review Letters* but referred on to the more specialized journals, because the results were not found to be of importance for the general physics community.

- Often an exiting new result has been made possible because a new method has been developed. In this case it is a good idea to publish the exiting new result first in a letter and then follow it with a full paper in which the new method is described in detail.

- *Note papers*: This is a short paper communicating a not-so-important result. Note papers are often concerned with technical issues and solutions – how to carry out a specific process more effectively, for example. Examples of Journals that include Note papers are *Precision Engineering* (*Technical Notes*) and *Journal of Vacuum Science and Technology B* (*Shop Notes*).

- *Review Papers*: These are longer, sometimes very long papers (up to around 50 pages or even more) presenting an overview of a given field, containing NO new results. Essentially this resembles a miniature text book within a given topic. Review papers are very important as an entrance for newcomers into a field. If you want to embark upon a new topic, the best thing that can happen to you is that you find a good review paper in that field. Review papers are often invited papers, which means that the editor of a particular journal asks an established scientist in a field to write a review. However you can also approach the editor of a journal and suggest a review topic. If you are a PhD student, it would normally be sensible to have your supervisor, as the established scientist, do this. A good argument for the necessity of a review paper on a given topic could be that a substantial part of the research has been published in conference proceedings not readily available through the *Web of Science* and/or in non-English speaking publications. This is still the case for many engineering topics that have only recently started to be published in "normal journals". Note that the structure of a review paper differs from that of other scientific papers because it is essentially a small textbook! Review papers will not be treated further in this book.

- *Technical Comment.* A technical comment (sometimes just referred to as a comment) is a text published in a journal, discussing a previous paper published in the same journal. The issue of discussion must be of "serious" nature, relating to serious mistakes/disagreements in the interpretation of results in the original paper or serious flaws in the experimental procedure.

- *Erratum/corrigendum.* This is a short text in a journal, essentially similar to a technical comment, but of a less "serious" nature. An erratum never challenges any key points in the original paper. A typical erratum would be the correction of a wrong formula, a wrong reference, etc.. An erratum is often written by the authors of the original paper. If you find a minor mistake in a recent paper, the courteous thing to do is to inform the authors of the original paper directly and suggest that they write the erratum themselves.

3.2. A Small Reflection on Technical Comments and Errata

Personally I consider technical comments and/or errata/corrigenda a very important part of the scientific process. It is in the nature of scientific work that we make errors and in the nature of scientific progress that these errors can be corrected by ourselves or by others. Further, modern publishing has made the system of technical comments and errata very efficient. If a particular paper has technical comments/errata there will be links to these at the electronic paper site and they will pop up in the publication databases. So I do encourage everybody to participate here and pay service to the scientific community - if you spent several hours pondering over a derivation before realizing that there is a printing mistake in the paper, do make the most of it by preventing it from happening to others.

Unfortunately many scientists are hesitant about writing technical comments on the papers of colleagues, basically because they are afraid of making enemies. Often they prefer to ignore errors completely or discuss them more discreetly, e.g., "sandwiched" in papers in which they present their own work. In many respects this is a pity because the information is then not easily traceable. Nonetheless it has to be said that the danger of making enemies is genuine. Not all scientists are altruistic and highly ethical and unfortunately, the universal search for truth which we (should) all subscribe to becomes less important to some than more immediate personal benefits and gains.

This is also the reason, no doubt, why some scientists are very hesitant to write technical comments on their own papers, because it seems to be embarrassing. This is clearly a bad attitude. The extreme attitude in the other direction I found in a couple of American colleagues: *"Well, perhaps we are sometimes a little bit fast in publishing in Physical Review Letters but then we just write a comment to our own publication, and then we have two publications in Physical Review Letters instead of one"*. I am not sure I really approve of this attitude either and it can backfire in the long run.

3.3. What Journal to Publish in?

Following the discussion in Section 2.3, it is clear that it is important where you choose to publish your paper. The most important rules are really quite simple and listed below. In addition there are some other factors you may consider. They are discussed in sections: 3.4, 3.5 and 3.6.

- To ensure that your paper reaches as many relevant readers as possible (so that it can be cited as much as possible), you want the journal itself to be as impressive as possible. Hence, you should publish your paper in the most relevant journal with the highest possible impact factor (see section 2.4). Normally (but not always) this will be a regular journal rather than conference proceedings. For a detailed discussion of conference proceedings see Section 3.6. If the choice is between a journal of impact factor 0.5 and 0.7, then it really does not matter which one you choose. But if the difference is between say 5 and 2, then it makes sense to try the journal with impact factor 5 first. You can always resubmit to a journal with a lower impact factor if you get rejected by your first choice. This is discussed in detail in Chapter 14.

    - The first thing you do when you have found some suitable journal candidates is to check out the impact factor of the journals. Usually the journals state their impact factors on their website, but alternatively you can directly to the *Journal Citation Reports* on the *Web of Science* website (see Section 2.4).

Chapter 3. The Journal Jungle

- **If a journal is not listed in the Web of Science, it does not have a proper impact factor and you should not publish in this journal.**
  - Do not get allured by flattering emails from journals encouraging you to publish with them. We all get those emails.
- The Time Argument: Rapid Publishing is an Issue.
  - In particular if you are a PhD student, you need to make sure that your work gets published reasonably quickly. Sometimes people worry that if they try the high-ranking journal first and get rejected, then they may not make it in time and that it is therefore better to "play safe" and go for a lower impact factor journal. I do not recommend this for the following reasons:
    * The refereeing procedure for the really high-ranking journals is usually very quick, so you do not waste too much time in trying. I have an ongoing bet with a colleague as to who can get the fastest rejection from *Nature*. At the moment he is leading: He has a paper which got rejected one hour after submission. As he dryly put it: *"It took Nature less time to reject my paper than it took me to fill out their online submission form"*. I should add to make the story complete that my colleague is a highly-respected scientist with many publications in high-ranking journals including *Nature* and *Science*, and that this was a very good paper which was later accepted by another not quite so high-ranking journal.
    * For the "middle-range" journals, with impact factors say between around 2 and 7, it is not always clear that it makes such a big difference where you try. There are several examples of people who tried the lower-ranking journals first, were rejected and then got published in the higher-ranking journals afterwards. The potential referees are the same for all these journals and the expectations of the referees for the journals in this range often do not differ much.
    * Many journals list an average processing time for their manuscripts. In the natural sciences this average processing time is usually 2-3 months, usually fast enough to allow enough time for at least one rejection (and resubmission) within a PhD period. If a journal keeps your paper for more than 2 months, you are entitled to push for an answer. Send an email to the editor.

- The Quality Argument: Is my work good enough?
    - I do not suggest that you try to publish every paper you write in *Nature* or *Science* that would clearly not be a professional approach and you might end up getting a bad reputation with the two journals. However, I do recommend that you retain a realistic optimism about the quality of your own work, "work your way down the ladder".
        * A famous professor once said to me: "*If I think the work is good enough for Physical Review Letters I try Nature. If I think it is good enough for Physical Review A, I try Physical Review Letters. The worst that can happen is that the paper is rejected and with a bit of luck I get some useful referee comments that can help improve the manuscript. Also, after more than 40 years as a scientist I am still often surprised about which of my papers that make it into the high ranking journals and which that do not.*
            · Remember! It is no shame to get a paper rejected. It happens quite often to most of us. See Chapter 14 for a discussion of how to handle the referee comments. Sometimes an initial rejection can be refuted.

3.4. Scientific Society Journals Versus Commercial Journals

As discussed in Section 1.1, the very first scientific journals were published by scientific societies. The journals published by scientific societies and institutions are in principle non-commercial since at least part of the profit will benefit the scientific society. Many such journals have high impact factors and high reputations. However, one must be a bit careful. The names "scientific society" or "institute" are not protected. Anyone can create a scientific society or institute and start publishing a journal, so the mere fact that a journal states that it is published by a scientific society is not enough. You should always check that it is a society of some standing in the scientific community with a reasonably large number of members and, of course, that the journal has an impact factor (see Section 2.4).

In addition to scientific societies and institutions, journals are published by commercial publishing houses. This includes some of the most prestigious journals such as all the *Nature Journals* but also many less prominent, yet still well-established jour-

Chapter 3. The Journal Jungle

nals. Scientific publishing is big business and big profit can be made. In 2010 the big, commercial publishing house *Elsevier* reported a 36% profit on its revenue amounting to more than 750 million euro. Universities and research institutes are secure and faithful customers! However you put it, this is a very large amount of money which has been generated by the scientific community and at the same time made at our cost, because the profit is made through our subscriptions. The situation really is a bit absurd: First the scientists do the scientific work and submit it (for free) for publication in a journal, then other scientists referee the work of their colleagues (for free), then the scientists who want to publish might have to pay a publication fee to the journal and finally the scientific institutions pay a subscription fee to access the journal.

There are many good *Elsevier* journals and I have published in *Elsevier* journals myself, but when I have a series of candidate journals I can choose between for publishing a paper and they all have roughly the same impact factor, I always try to publish in journals published by well-established scientific societies rather than purely commercial journals. To complicate matters, there are several cases of scientific societies outsourcing the publication of their journals to commercial publishing houses.

3.5. Open Access Journals

Initially all scientific journals were published in a printed version only and distributed to the subscribers (university libraries, university departments, research institutes, private subscribers, etc.) on a regular basis. If you wanted to get hold of a paper published in a journal, you would have to find the journal in the library and then make a photocopy, which was time consuming. If your local library did not hold a subscription, you would have to order a copy from another library, which was even more time-consuming and expensive.

The internet has revolutionized the access to journals. Nowadays essentially all journals in the natural sciences are available online and can be downloaded directly if you have a subscription key or make a credit card payment. This is truly a revolution. The next, natural step in this development is to eliminate the printed journal altogether (fine by me, I cannot remember the last time I held the paper-copy of a journal in my hand!). This reduces the price of publishing immensely, since the whole printing process can now be avoided. This development has led to the instigation of open access journals. Open access journals are journals that are usually only available online and do not require any subscription: they are freely available to everybody - hence the name "open access". The journals finance themselves through publication

3.5. Open Access Journals

fees paid by the authors. Over the last few years, the number of open access journals has skyrocketed and hardly a day goes by in which I do not receive an email invitation to publish in or be on the editorial board of some obscure new journal with a very scientific sounding name.

There has been a lot of political pressure for increasing open access publication. On 17 July 2012, the *European Commission* issued a press release stating, "*The Commission will make open access to scientific publications a general principle of Horizon 2020, the EU's Research & Innovation funding program for 2014-2020. As of 2014, all articles produced with funding from Horizon 2020 will have to be accessible*". The *Global Research Council* has announced that they will discuss a global policy for open access. Some of the established journals try to handle this by offering to make an article published in a subscription journal available as open access for an increased publication fee (more money off the scientific community).

Nature Publishing Group has started two open access journals: *Nature Communications* and *Scientific Reports*. *Science* has recently followed suit with the open access journal *Science Advances*.

In principle open access is clearly a good thing and in some cases it can lead to a degree of dissemination that would otherwise not have been possible. I experienced this recently myself. In 2012 together with colleagues I published an article in the open access journal *Scientific Reports* mentioned above. The paper presents an investigation of a 2800 year old plant fiber textile found in a Danish Burial Mound [18] and is the result of a cooperation with colleagues from Archeology and Geology. By combining various methods, we were able to show that the textile is made of nettle fibers and that, very surprisingly, these nettle fibers do not come from Denmark. This indicates that nettle played a much more important role in Bronze Age Europe than what had hitherto been assumed. The paper was selected as a Science Editors Highlight [19]. In addition the paper received some publicity in the media and it has now been viewed more than 7000 times, which I am pretty sure is considerably more than almost any other paper I have ever contributed to. I am also sure that a lot of the people who viewed this paper were "interested public" rather than active scientists, and this to me is a very nice thought. Directly or indirectly my work is largely financed by tax-payers and it is very nice to be able to give something back immediately.

The one crucial problem with open access is quality control, as discussed for example in a recent article in *Science*. Here John Bohannon describes how he managed to get a spoof paper accepted for publication in more than 150 open access journals [20]. In Section 2.8.1, I discuss the upcoming danger of faked impact factors. As scientists we can be expected to be able to access the quality of the journals and papers within our

Chapter 3. The Journal Jungle

field, but we cannot expect the interested public, including journalists and politicians, to be able to do so. Hence, there is a danger that in due course, low quality or deliberately misleading work in obscure, open access journals will diminish public faith in science.

Nobody can or should be prevented from starting a journal, but in my opinion the *Global Research Council* and the *European Union* have the obligation to make sure that their welcome push for open access does not backfire into open confusion or even open misleading.

3.6. Conference Proceedings Versus Regular Journals

Attending scientific conferences is a very important aspect of scientific life. Conferences are the places where you present your work directly to your colleagues and enter into scientific discussions. Conferences are related to scientific paper writing through the concepts "conference abstract" and "conference proceedings paper". Let's take a look at these two concepts.

3.6.1. Conference Abstract

When you want to give a presentation at a conference and you are not an invited speaker, you have to submit an abstract or alternatively what is usually termed "a conference paper". Based on this abstract/paper, it will be decided if you work is accepted for the conference and whether you will be offered an oral presentation (a talk) or a poster presentation. In some fields it can be very difficult just to get accepted for certain conferences. In other fields (such as my own) it is usually not a problem to get accepted for the conference with a poster presentation, but getting to do an oral presentation is quite competitive. You should always try to get an oral presentation (see Section 10.2).

- To confuse matters, "conference abstract" can mean different things.
  - Short abstract (regular abstract), typically 200-500 words: In this case you should write your conference abstract just the way you would write a normal abstract for a journal paper (see Section 5.3), with the small difference that you include citations in the conference abstract. **It does not harm to check who is on the organizing committee for the conference and make sure you do not forget to cite their work (if relevant of course).**
  - Extended abstract (more than one page and typically up to 4 pages): Extended abstracts sometimes confuse people because they think they are supposed to take the word abstract literally and write the extended abstract the way they would write a regular abstract. This often leads to a very long introductory text before the actual results are presented. If the referees have to read very many abstracts, this can produce negative effects. To make life easier for the referee, write the extended conference abstract like a mini paper, beginning with a short abstract (in italics) written in the usual way (see Section 5.3) followed by the usual sections (see Chapter 5).

3.6.2. Conference Proceedings Paper

Many conferences offer the possibility of submitting a paper for the conference proceedings. In some cases you have to submit a paper to be accepted for the conference, as discussed above. In other cases you have the option of submitting a paper before a certain deadline, once your abstract has been accepted for the conference.

- There is no difference between how a conference proceedings paper and a journal paper should be structured.
  - In some fields conference proceedings can be very prestigious. In some fields within Computer Science, for example, there are conferences where

the acceptance rate is comparable to the acceptance rate for *Science* and *Nature*. The conference proceedings for these conferences are published annually and are available in the *Web of Science*; they have high impact factors.

– In other fields, including my own, conference proceedings are not considered prestigious. To encourage people to submit work for the conference proceedings, they are often published as a special issue of an established journal.

* **Be careful not to "waste" your exciting results on low-ranking conference proceedings.**

* As a general rule, once you have published your results in a conference proceedings paper, you cannot publish the same results elsewhere without making a reference to the conference proceedings, but this is all a bit of a gray area. For example, abstracts are generally not a problem and do need to be cited in a later paper even if a long abstract is essentially a short paper. Only if your text is officially labeled as a "conference proceedings paper", do you need to cite it in future publications. An exception to this can even be made if the conference proceedings volume does not have an International Standard Book Number (ISBN) or an International Standard Serial Number (ISSN). Then it does not count as a proper publication, but rather as a private communication within a closed circle, and you can publish the same results (the same paper even) without any reference to the first publication.

* It is possible to present the same scientific results at several conferences without being at all obliged to mention that you have already presented some results elsewhere. In fact it can be a very good idea to present the same results in several places to ensure that you reach all the relevant scientific communities. If you do so, I recommend that you change the abstract text and in particular the abstract title for the different conferences and adapt it to each conference (also makes your CV look more impressive), but you do not have to.

* **Be careful not to waste your time by going to conferences which do not have a good standing in your scientific community.** There are quite a lot of conferences out there which may have seriously sounding names but which are essentially just an excuse

for the organizers to make money and the participants to have a free holiday. For an amusing blog on "bogus" conferences see [21].

Whatever you do, make sure that your results are published in journals accessible through the *Web of Science*. This is important both for your own sake, because the official Hirsch index is calculated on the basis of journals that are accessible in the *Web of Science*, but also because this is the best way to ensure that your results are available to a broad scientific community. If your results are not published in a *Web of Science* journal, it will be very difficult for people to find this journal, in particular if they are not exactly in the same field as you and do not have the full overview of all relevant journals.

3.7. Publishing Interdisciplinary Work

Interdisciplinary work can be extremely interesting and rewarding, but it represents a particular challenge when it comes to publishing, because you are addressing several, different scientific communities and they will normally be interested in different aspects of your work. Also these different communities will probably not read the same journals. The best solution to this dilemma is to publish essentially the same results in different journals but emphasize different aspects. It is usually not possible to publish the same results twice, so the emphasis is definitely on the different aspects. Transparency must be a top priority here; previous papers with previously published results must be properly referenced. Also, if you wish to publish in a really high-ranking journal, you should first submit there.

Interdisciplinary work combining natural science and/or engineering with the humanities and arts presents, in a sense, the most extreme cases of interdisciplinary work. There are not many journals open for this type of work, but there are a few: *Nature* and *Science* of course, but also *Proceedings of the National Academy of Sciences (PNAS)*, which is almost as prestigious. In addition there is *Nature Publishing Group*'s new open access journal *Scientific Reports*, which is open to all disciplines. *Interdisciplinary Science Review*, published by the *Institute of Materials, Minerals and Mining*, London, is another interesting journal devoted to the interrelation of science and technology with the humanities and arts.

3.8. Maximize the "Findability" of Your Papers

3.8.1. Publication Search Engines/Databases

If people cannot find your paper, they cannot read it and if they cannot read it, they cannot cite it. "Findability" is very important! For this reason alone it is crucial to publish in a journal with an official impact factor (see Section 2.4) because your paper will then appear automatically in the *Web of Science* publication search engine. However, the *Web of Science* is not the only publication search engine available and it has the disadvantage of being only available for a fee, so not everybody has access to it everywhere. It is therefore important to check that your paper also appears in other (free) publication search engines and that it is registered correctly everywhere. Everything works automatically, but it is better to check. The most important publication search engine in the life sciences is *PubMed*. It is free for everybody. It was developed and is maintained by the *National Center for Biotechnology Information (NCBI)* at the *U.S. National Library of Medicines (NLM)*, located at the *National Institutes of Health (NIH)*. *PubMed* searches the *NLM* bibliographic database *MEDLINE*. Journals are included in *MEDLINE* based on the recommendation of the *Literature Selection Technical Review Committee* appointed by the *NIH*. In other words, *PubMed* has quality control. For more details see [22]. *Google Scholar* is another search engine for all sciences and it is becoming increasingly popular. As the name suggests, it is maintained by *Google* and it has free access. A crucial difference compared to the other search engines discussed here is that *Google Scholar* does not have any quality control. It uses the whole internet as its database and it includes much more than proper papers, i.e. random conference and seminar presentations that happen to have been published on the internet as well as obscure open access journals. Hence it can be both an advantage and a disadvantage to use *Google Scholar*. At present there is no freely available, quality-controlled publication search engine for all scientific disciplines. This is unfortunate, not least given the large increase in the number of journals over the last few years (see Section 3.5).

Remember that if people are interested in one of your papers, they might be interested in some of your other papers as well. Therefore it is important that it is easy for people to find you as an author when using the search engines. This is discussed in Section 5.2.5.

3.8.2. ResearchGate and Other Online Social Networks

Online social networks can be a way of increasing the visibility as well as the "findability" of your work. If you do not already have a permanent position it is particularly important to appear reasonably serious and professional and therefore it may be sensible to separate your private network activities and your professional network activities rather strictly.

Personally I have found that from a "findability" perspective *ResearchGate* is quite useful. *ResearchGate* was started in 2008 by virologist and computer scientist Dr. Ijad Madisch [23] and now has nominally around 7 million users, though it is not clear how many of these are really active. In *ResearchGate* you can create a profile and upload information about your publications, and copyright permitting, the publication itself for direct downloading. *ResearchGate* does an automatic copyright check for you, but it remains your legal responsibility. In any case, since many journals provide statistics on how many times a paper has been downloaded from their website, and you may want to boast in your CV about your paper being among the most downloaded papers in a given journal, it might actually be more clever just to provide the link to the original journal website for downloading, similar to how it is done by *Web of Science*. When you set up your profile, *ResearchGate* automatically "crawls" the net to find all your publications. Do check for completeness and redundancy.

In principle you can search for an author with any journal search engine, but *ResearchGate* is very easy to use; it is free of charge and you can upload material not published elsewhere, write comments to your own papers and so on. *ResearchGate* also has nice features, such as the possibility of "following" other scientists so that you are notified automatically if they upload new publications.

ResearchGate has been exposed to some criticism: It is accused (rightfully) of sending too many automatically generated emails, of not being fully transparent as to how the so-called RG score, which is assigned to each profile holder, is generated. Finally you should bear in mind when setting up your profile that though joining is free of charge, *ResearchGate* is a private enterprise and you do not know how the information that you provide is used.

Google Scholar mentioned in the paragraph above also provides the option for you to create your own personal profile listing your publications.

Chapter 3. The Journal Jungle

3.9. *Nature* versus *Science* - A PHD Comics Story

NATURE vs. SCIENCE

FOUNDED:	1869	1880
Published by:	Nature Publishing Group (a division of MacMillan Publishers Ltd. of London, a subsidiary of Verlagsgruppe Georg Von Holtzbrinck, GmbH)	American Association for the Advancement of Science (AAAS)
Cost:	£10	$10
Impact Factor:	31.434	28.103
	(It is important to compute this to the third decimal. Units: inches)	
Sections:	News News Features Correspondence Perspectives Articles Letters Jobs To-*mah*-toe	News of the Week News Focus Letters Views Research Articles Reports Careers To*ma*to
Ads per issue:		
Full page ads:	16	9
Full page ads about itself:	6	5
Full page ads featuring people in white lab coats smiling and pipetting something:	5	4
Which one will you submit your paper to?	If only you had that problem.	

JORGE CHAM © 2009

WWW.PHDCOMICS.COM

3.9. Nature versus Science - A PHD Comics Story

Chapter 3. The Journal Jungle

3.9. *Nature* versus *Science* - A PHD Comics Story

Part II.
Writing the Paper

Chapter 4.
Getting Started

This may sound a bit odd, but getting started with the paper is often the hardest part of the paper writing process. When I say getting started with the paper, I mean the step where the main emphasis of your work moves from "doing" to "presenting". This is moving from actually doing the scientific work in the lab/field/office/archives to writing about it. It includes polishing/performing the data analysis, making the figures, etc.. Ideally, a certain amount of writing and data analysis should be performed while the actual scientific work is being done, but in practice this is often not the case. Hence, the transition from getting the results to writing about them is often quite abrupt.

4.1. Why is it so Difficult to Get Started?

There are several reasons for why it can be so hard to move from "doing" to "presenting". I'll list a couple here, with some suggestions for how you can try to overcome them. However, as I know from deep, personal experience, there are no miracle solutions.

- The time factor.
    - Good papers in well-acknowledged peer-reviewed journals is the most important tool you have for promoting your scientific career as discussed in Section 2.1, but unlike other aspects of scientific life, i.e. lecturing, supervising, writing grant applications, reporting on running grants, submitting abstracts for and attending conferences, giving talks, doing obscure administrative and political work required by the university, etc., all of which will be associated with deadlines, there is no immediate time pressure for paper writing. You can submit a paper to *Nature* whenever you want. This can actually be a serious problem, because it means that paper writing, even

4.1. Why is it so Difficult to Get Started?

though it is arguably the most important thing you have to do, is often given the lowest priority on a day-to-day basis.

- Possible solutions:

    * **The first, most important thing is to allocate regular, fixed, long time slots for writing**, ideally on a weekly basis. If you are in a position where you are likely to be disturbed a lot during the day, do the writing at home because paper writing, in particular thinking about the results, requires periods of uninterrupted concentration. In my experience, when writing a paper I need at least half a day to get anything productive done and a full day is better.

        · **During these time slots make sure that your email, Skype, Twitter, Facebook, LinkedIn, mobile phone, etc. are off!** If you have problems with this, you already suffer from addiction! You should only be accessible for emergency telephone calls (I repeat - no facebook, no emails!).

    * **The second, most important thing (which is just as important as the first!) is to allocate regular, fixed time slots on a weekly basis for unforeseen work and work that takes longer than you thought it would.** If you do not allocate these extra time slots, what will inevitably happen is that your paper writing time slots will be eaten up.

    * If you cannot find time in your calendar for these two types of time slots, then your schedule is simply too full, you are involved in too many things and you must try to reduce the overall workload. This is much easier said than done, I realize.

- The fear of discovering that your results are in some sense wrong or that some results are missing when you sit down to write.

    - This may be a problem which applies especially to experimental work. It is quite normal in many areas of experimental science to work very hard in the lab for a while, not doing data analysis at the same time. This is not ideal, but in reality this is what often happens. You are then faced with the situation that when you finally get down to doing the data analysis, your experimental setup may no longer be up and running, which means that it will be difficult to repeat/extend the experiments. Computer simulations requiring very long simulation times fall into this category as well.

Chapter 4. Getting Started

- As you have probably gathered, there is no good solution for this. The best is clearly to avoid the situation, but if you have gotten yourself into it, the only thing to do is to be honest with yourself. Often people do not really want to admit to themselves or others, that this is what they are really afraid of *"I have other important things to do."* is a less embarrassing explanation. However, if you are not honest with yourself and bite the bullet, you risk procrastinating for months and feeling bad, sometimes for no reason at all. Usually it is possible to put together a decent paper even if the data-set is not quite complete. As my PhD supervisor put it to me during my PhD: *"There is no such thing as a perfect experiment"*.

- You love doing scientific work, but you do not like writing papers.

    - I am not quite sure, but I suspect that this is also a problem that is most prominent in the experimental sciences. I have met several people who were brilliant in the laboratory, but who did not like writing. Indeed the process of writing requires very different skills from the ones you need in the lab.

    - Possible solutions:

        * I hope this book can help you. More specifically I suggest that you try to think carefully about what aspect of the writing process it is

4.1. Why is it so Difficult to Get Started?

that you do not like or have problems with: Are you uncomfortable writing in English? Do you have problems structuring your results and extracting the take-home message? (see Section 4.2) Do you not like making figures? (Personally I hate making figures!) A careful reading of Section 6 on how to draft the manuscript may help you to pinpoint your trouble spots and there are suggestions for how to overcome them.

* Structure the writing process so that you work actively together with the co-authors (perhaps getting them to do the parts you particularly dislike). Active involvement of co-authors also increases the pressure to finish the paper.

- In principle you like writing papers, but sometimes you find it very difficult to concentrate on the writing process.

  - Do not expect miracles from yourself. Your brain is part of your body and your body cannot work equally well every day. I confess that this is something I had trouble with myself for a long time and I have wasted a lot of time trying to force my brain to work. In fact I had to be taught a lesson from my husband who practiced high-level athletics in his youth. He told me how athletes experience the limits of their bodies all the time. A very important aspect of the training process is to time it so that the performance peaks at the major sporting events, but sometimes you are hit by a bad day, and then you lose regardless. The same phenomenon can be found with all performing artists. They have to live with the fact that they will not always be able to perform to their best in front of an audience. In comparison to athletes and performing artists, we scientists actually have it easy, because we do not have to peak at specific times. Having a bad day does not mean that you should drop your allocated paper writing time slots. In fact it is very important that you stick to them! Just spend the time doing less demanding tasks, such as cleaning up the references, writing the methods section, etc.. If you are lucky you may even find that your brain relaxes during this process to such an extent that you can tackle the more demanding work after some time.

Chapter 4. Getting Started

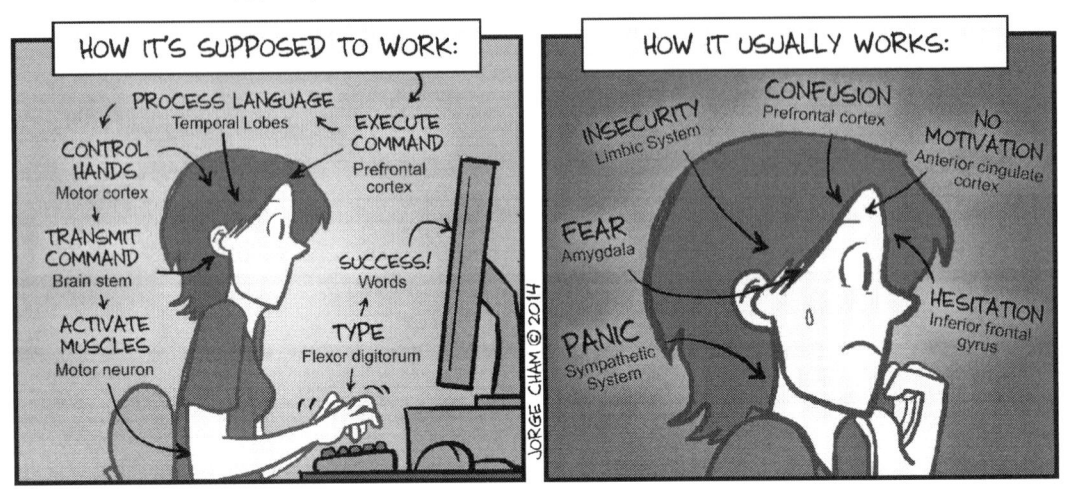

4.1.1. A Small Reflection on (Experimental) Scientific Work

Most professors heading experimental groups are rarely seen in the labs, because they have to spend their time with other tasks such as writing/correcting scientific papers, writing grant applications (experimental work is often very expensive), managing all the administrative aspects of running an experimental facility and "selling" the results at conferences and seminars. It is simply too much work for one person to do all this and the experimental work at the same time. Hence the experimental work is done by students and postdocs, and not always in a very efficient way because they lack good, practical supervision in the laboratory. Sometimes this leads to very interesting new approaches. It is important to remember that part of a PhD education is learning to learn on your own and being able to attack problems for which you do not know from the beginning how to achieve the solution (and cope mentally with the situations where there are no solutions). This inevitably means wasting time in a good sense, but unfortunately there is also a lot of unnecessary time wasted in academia due to lack of supervision.

4.1. Why is it so Difficult to Get Started?

In a way it is absurd, but if your great skill and interest is experimental lab work, becoming a professor may not be your most satisfying choice of career, or even worse, it may be difficult for you to even get a chance to become a professor because you lack the skills for and interest in selling yourself and your ideas which is so important in modern academic life.

I have no solution for this other than that you read this book and practice. Remember that paper writing is a skill that can be learned! Nonetheless there seems to be a flaw in the present academic system. My observation has been that often the most productive research groups are the ones who have managed to join three people with complimentary skills. One person who is good at writing and selling and getting new ideas, one person who is good at realizing these ideas experimentally and one person who is good at managing all the daily, practical matters.

Chapter 4. Getting Started

4.2. General Tips Before You Start Writing - What is the "Take-Home Message" of Your Paper?

- **Before you start writing, decide what should be the "take-home message" of your paper.** There should always be one clear, main take-home message (one main result) and perhaps, but not necessarily, a few additional minor results.

- Think about how people search for papers. Today in the natural sciences almost all paper search is done via the internet as a search for keywords in title and abstract.

- **Scientific papers are read on many different levels:** People will start by skimming over the title. If the title sounds interesting, they might read the abstract, which is usually available on the net even if you haven´t paid the electronic subscription. Then they will typically look at the figures. This is one of the reasons why figure captions are very important (see Section 5.8). Following that, they will look at the conclusion. Eventually they might read the whole paper and then they might cite it: This is your goal!

Chapter 4. *Getting Started*

- If the readers can get all the information they need for citing your paper just by reading the abstract, so much the better.

- The reader should understand what the key message, the take-home message of your paper is on all of the levels mentioned above! Do not be afraid of repeating yourself.

- A paper should tell a concise story. This story should be largely clear to you before you start writing.

4.3. Pitfalls

- You paper is not a diary describing your research chronologically.

    - Focus on the final success: Often you try several approaches that do not work before you finally find something that sort of works; then you may get all results within a relatively short time. You should write the paper as if your final approach was your initial approach. Sometimes people feel that this is cheating or making it seem as if you have done less than you actually did. However, for the reader, it would be very confusing to read a diary-like account. Remember the main purpose of the paper is that the reader understands what your take-home message is (see Section 4.2). A simple, clear explanation of the final development is easier to follow and, if anything, it will just impress people more.

- **It does not matter what your initial goal was, what matters is where you ended up – which result you got in the end. Or to put it differently: You may have thought of an initial "take-home message", but it evolved into a completely different "take-home message".**

    - Not infrequently do you end up with a result which is completely different from what you intended to investigate/measure in the first place. Often these results can be the most interesting. One of the most famous examples of this is X-rays. Röngten's initial aim was to investigate the light emission stemming from discharges in a novel sort of vacuum tube. He did not expect to make a discovery which led to one of the most revolutionary tools in solid state physics, material science, structural biology and modern medicine. **Writing your paper as if the results you are presenting were what you aimed for all the time is NOT cheating.** Stylistically it is much easier for the reader to follow if you write it that way. Remember the main purpose of the paper is that the reader understands what your take-home message is (see Section 4.2), NOT what prompted you to do this work in the first place.

- **The reader does not care about your failed attempts, only the final success.** Negative results are difficult to publish. By negative results, I mean for example, a test for an effect which was not found, an attempt to make a process work which did not succeed, etc.. Some journals dedicated to negative results do exist, but they are not accepted as impact factor journals (see Section 2.4).
    - Some exceptions to the rule that you cannot publish negative results can be found. If, for example, you are testing an established hypothesis and manage to form your negative result into a positive results with an "upper limit argument".
        * Example 1: Try to detect a change in the gravitational constant when it acts over small distances - negative result, no change found. Positive result with "upper limit argument": "*The gravitational constant remains constant to within a precision of X over distances down to Y nm*" - where X and Y are determined by the statistics of the experiments.
        * Example 2: Test a drug for an expected effect or side effect - negative result, no effect or side effect found. Positive result with "upper limit argument": "*The drug had an effect of less than X%*" - where X% is determined by the statistics of the experiments.
    - A drawback of the reluctance to publish negative results is that no doubt, many people try independently to do the same work and fail. If you want to help your scientific colleagues, it can be very good to include a (brief) discussion of previous failed attempts in the methods section of a paper (see Section 5.6). Negative results can also be "smuggled" into a paper as minor results following a major "take-home message".

4.3. Pitfalls

- Sometimes your problem is not that you have too little data but that you have too much data. If you cannot find one, single, clear "take-home message" for your data, perhaps that is because you do in fact have two equally important "take-home messages" and you need to write two papers instead of one!

- It does not make you look better as a scientist if your papers are very long. Papers with a single, clear take-home message are easier to read.

Chapter 5.
The Structure of a Scientific Paper

A typical section structure for a scientific paper is listed below. It should not be taken as completely fixed, however all the issues listed below (except "*Acknowledgments*" and "*Appendices*") must ALWAYS be addressed in a paper. Note the following points:

- Some journals require a pre-set structure that may differ slightly from the one listed here. For example, some journals require a special "*state-of-the-art*" section, whereas in other journals the "*state-of-the-art*" is included in the introduction section.

- Some of the points listed below will often be joined together, for example the "*Discussion*" section is often joined with either "*Results and Analysis*" or "*Conclusion*". This is really a matter of taste.

- If the journal you want to publish in does not require a preset structure, then the structure below provides a good starting point. **Note, it is usually possible to include sub-sections or additional sections in between.**

    - In a recent paper from my group we included an extra section called "*Atom-sieve Fabrication*" after "*Experimental Setup*" [24]. This was a useful way to separate two very different areas of technical information, related to the testing of the atom sieve with an atomic beam, and the fabrication of the atom sieve itself with electron lithography.

- Note that **letter style papers (see Section 3.1) will often not have any section headings, just paragraphs. The structure below should still be followed.**

In this chapter we will go through the context of each of the sections listed below step by step. The greatest emphasis will be put on the first three sections, since they are firstly very field independent and secondly, they are the part of the paper accessible to everybody. They do not require a subscription to the journal, so they are particularly important as "advertising" material.

The different sections of a paper are usually NOT written in the order listed below. The whole of Chapter 6 is devoted to a discussion on how to structure the actual writing of the paper in the best way. In the present chapter, we just discuss the context of each section. It may actually be helpful if you read Chapter 6 before this chapter.

- Title
- Authors
- Abstract
- Keywords and/or Subject Classification Numbers
- Introduction
- Methods/Experimental Setup
- Theoretical Background
- Results and Analysis
- Discussion
- Conclusion
- Acknowledgments
- Citations
- Appendices (online supplementary material)

Chapter 5. The Structure of a Scientific Paper

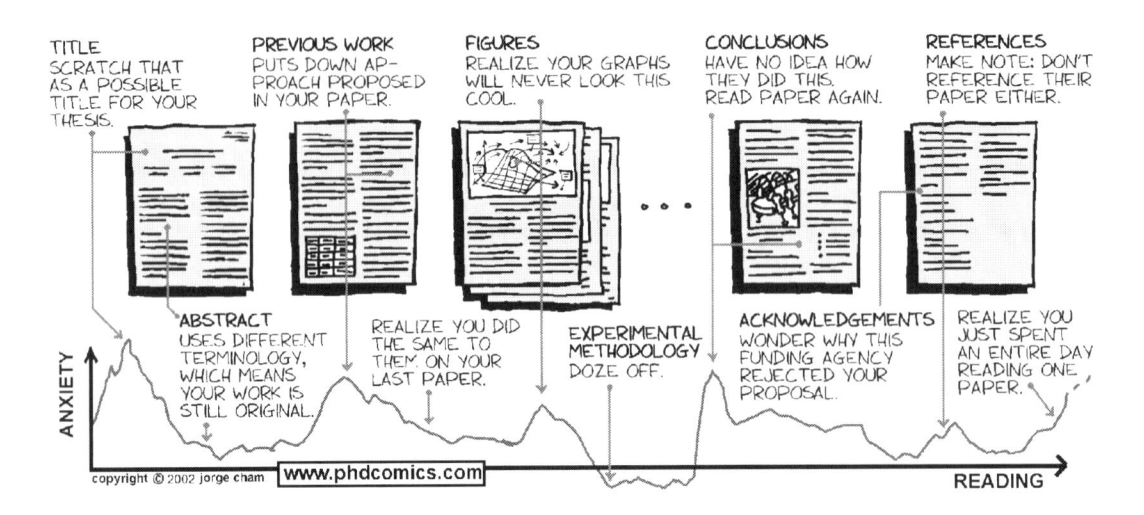

5.1. Title

The title is crucial because it determines to a large extent if people will actually read your paper. It is the first part of your paper that people will see, regardless of whether they are doing a regular browsing of selected journals or your paper has popped up as a result of a specific search in a search engine.

You should adhere to the following rules:

- Spell out all words completely in the title. Do not use abbreviations (i.e.: Atomic Force Microscopy rather than AFM). There are many papers out there with abbreviations in the title, but it is not so nice for newcomers in the field. However, make sure that you introduce common abbreviations (AFM) in the abstract. Some people only do keyword searches on the abbreviations.

    - The exception to this rule is established acronyms for large scale research collaborations, of which there are many.

        * Examples: ALICE = A Large Ion Collider Experiment, AEgIS = Antimatter Experiment: Gravity, Interferometry, Spectroscopy, etc..

        * For these cases, I recommend that you introduce the meaning of the abbreviation in the abstract or at the very least in the main paper, but many people do not.

5.1. Title

Clever Acronyms: the Holy Grail of Academia

Step 1: Use the loose definition of the word "acronym"

Step 3: Missing a letter? Pull out an obscure buzzword that fits!

Step 4: Desperate? Just pick letters from the middle. I'm sure no one will notice.

ACtually Random Onomastic iNitials You Make (up)

Step 2: Is it coherent? Does it makes sense? What matters is that it *sounds* cool.

Step 5: Ignore words that don't contribute. Kind of like your part in the project.

Types of Acronyms:

- **Folksy Names:** a cheery name will distract people from the fact your project cost millions — A.L.I.C.E., B.O.B., D.A.V.E. ✓   A.D.O.L.F., Z.I.P.P.O., S.I.G.M.U.N.D. ✗

- **Aggressive verb/predatory animal:** a requirement for getting military funding — K.I.L.L., S.H.A.R.K., W.O.L.F. ✓   O.B.L.I.T.E.R.A.T.E. (too many words!), B.U.N.N.Y. ✗

- **Greek names:** nothing says "Sci-Fi" like a good greek name — O.M.E.G.A., A.L.P.H.A., S.I.R.I.U.S. ✓   T.O.G.A., P.I.T.A., T.Z.A.T.Z.I.K.I. ✗

Remember: Acronyms cleverly reveal one's nimble youthful mastery abbreviating construed rigidly opted nomenclature, yielding monetary awards contracting research overtures not yet manifested!

Bonus points: make your acronym recursive!

JORGE CHAM © 2008

WWW.PHDCOMICS.COM

- The title should be short but long enough to accurately describe what the paper is about. This means:

    - The title must be general enough to ensure that you reach a broad readership.

    - The title must be specific enough so that you do not „drown in the crowd".

        * Make sure that key-words for the work that you do not use in the title are mentioned in the abstract.

        * Tip: The PhD student test: Would a first-year PhD student who has just started working in your general area find your paper if he/she was interested in the topic?

- Additional tip: Scientists are supposed to be specialists but it is not good to appear limited or narrow-minded: Try to vary the titles of your papers in a way

55

that they reflect your flexibility with different topics and techniques even though the basic approach might be quite similar.

- I have worked with molecular beams all my scientific life, so I could have more than 30 papers with the title: "*A molecular beam investigation of A*", "*A molecular beam investigation of B*", "*A molecular beam investigation of C*", and so on. The titles would be perfectly adequate in terms of explaining what I was working on, but a list like that might create the impression that I am not a very interesting and creative scientist, which is of course not the case.

- The "*declarative title*" described in the next section, is a particularly useful tool for creating variety in your titles.

5.1.1. Declarative Versus Neutral Titles (This is Important)

You can make your title more interesting by presenting the main result of the paper, "the take-home message", in the title rather than just using the title to describe the topic/area of investigation in more general terms. This is what Björn Gustavii [25] refers to as the declarative versus the neutral title. To illustrate this, here are some titles selected from *Nature* and the *Journal of Food Engineering*:

- Titles from *Nature*
  - *How long is a giant sperm?* [26]
    * This is a declarative title. It tells you that the main result of this paper ("the take-home message") is to answer the question: How long is a giant sperm? Note the curiosity effect of the question mark. Titles with question marks can be very nice. For an amusing reflection on titles with question marks versus other type of titles see [27]
  - *Graphene-based composite materials* [28]
    * This is a neutral title just stating the general topic covered by the paper (in fact it is a review paper).
  - *Kruppel-like factor 2 regulates thymocyte and T-cell migration* [29]
  - *An RNA-dependent RNA polymerase is required for paramutation in maize* [30]

5.1. Title

* These last two titles are both declarative titles. Note how both titles are formulated as scientific statements: **The titles are the take-home messages of the papers**. You do not even need to read the abstracts to get the take-home message. Even to an uninformed reader like me it is completely clear what the take-home message is, though I do not understand what it actually means or why it is so important.

- Titles from the *Journal of Food Engineering*
  - *NMR imaging of continuous and intermittent drying of pasta* [31]
    * This is a neutral title, stating the general topic area. A declarative title would present an actual result related to the drying of pasta.
  - *Effect of dietary fiber addition on the selected nutritional properties of cookies* [32]
    * Again, this is a neutral title; it does not provide any direct information about the effect of dietary fiber addition to cookies.
  - *Mathematical modeling of mass transfer in osmotic dehydration of onion slices* [33]
    * Again, this is a neutral title. It does not provide any direct information about the osmotic dehydration of onion slices.

Nature is one of the highest ranking scientific journals, aimed at a general scientific audience whereas the *Journal of Food Engineering* is a very specialized journal aimed at a much smaller expert community. On average the papers presented in *Nature* will therefore contain results that are more exciting than the results presented in the *Journal of Food Engineering*. Hence, it is not surprising that *Nature* may have more papers with declarative titles than the *Journal of Food Engineering*. Not all papers are suited for having a declarative title and that is perfectly fine. However, if the "take-home message" of a paper is suited for a declarative title, then it can make the paper sound much more interesting. It is also a good way to let your titles appear more varied (as discussed in the previous section). To illustrate this, here is an example from my own work:

- Actual title (declarative title, the title is the take-home message)
  - *Observation of the boson peak at the surface of vitreous silica* [34]

Chapter 5. The Structure of a Scientific Paper

- A neutral title would have been
  - *A molecular beam investigation of dynamic properties of the vitreous silica surface*

5.1.2. Two Sentence Titles

Think about the possibility of having a "two sentence" title. A catchy "first title" with a more explanatory subtitle. This often works well regardless of whether you want to have a declarative or a neutral title. Some examples are listed below:

- *Imaging with neutral atoms: a new matter wave microscope* [35]
  - This is one of my own papers. The first bit is supposed to be the catchy bit, and I think it does sound catchy (at least if you are working with microscopes).

- *Can antibiotics make bacteria live longer? An investigation of ...*
  - This is a paper by a former workshop participant. She had discovered that a certain strand of bacteria was not only not killed by a certain type of antibiotics, the bacteria ate the antibiotics as food. The second part of the title was then the technical explanation.

- *Shaken not stirred: On the influence of agitated liquids on the formation of bio-films in ...*
  - This again is the title of a paper by a workshop participant. The beauty is that shaken was actually better in this context. Unfortunately it must have been a bit too radical for the supervisor, because I have not found it in the literature.

5.1.3. "Empty" Words in Titles

Words like „*investigation of*" or „*observations of*", "*measurement of*" are in a sense "empty" because it is clear from the beginning that you have performed an investigation or observation. Some people are very keen on using them in the title (indeed I have done so myself on several occasions), and you can do that if you wish to. Sometimes it can even be necessary, particularly if you want to emphasize that your work is experimental, then you may need "*measurement of*". However, your title may sound more catchy and snappy (and it will be shorter) if you try to avoid these words. Below an example (made up for this occasion, just in case you should wonder).

- Instead of writing:
  - *An investigation of the influence of color on the taste of apples*
- Consider writing (removing "empty" words):
  - *The influence of color on the taste of apples*
  - alternatively: *"Is the taste of apples influenced by their color?"*

5.1.4. Using "Dynamic" Rather than "Static" Language

This is a minor point, but sometimes you can consider making your title sound more "lively" by using dynamic rather than static language. Example:

- Instead of writing (static language)
  - *Tuning the paint deposition conditions for making green apples red*
- Consider writing (dynamic language)
  - *How to tune the paint deposition conditions for making green apples red*

5.1.5. Humorous Titles

Some scientists are not so keen on humorous titles, which they see as being non-serious or even inappropriate. Before you use a humorous title, it is therefore better first to check the general trend of titles in the journal in which you want to publish. **Needless to say, humorous titles are only appropriate if you are working on non-sensitive topics.** I do not recommend the use of humorous titles if you do research on cancer, obesity, etc.. Sensitive research-topics apart, using humorous titles is a good thing because it makes people pay attention. A humorous title for a conference paper can mean that people go to your talk instead of somebody else's. *Nature, Science,* etc. are big users of catchy, sometimes humorous titles. When *Nature* published a research highlight mentioning one of the papers from my group "*Poisson's spot with molecules*" (another example where the title is the take-home message) [36], their highlight title was: "*Atom optics - seeing spots*". The first author, my PhD student Thomas Reisinger, was not entirely pleased, but we agreed that in comparison to one of the other highlights that week: "*Dude, where is my dot?*", we had actually been quite lucky [37]. There is a legend in the physics community about a bet as to how to manage to sneak the word "penguin" into a

Chapter 5. The Structure of a Scientific Paper

Physical Review Letter title, *Physical Review Letter* being one of the most prestigious journals in Physics. While writing this paper I researched the story and found a nice account of it [38]. It turned out that the bet (which was made in a pub and related to a game of darts) was actually a bit more modest. The bet (between Melissa Franklin and John Ellis) was just to get the word "penguin" into a physics paper. John Ellis won the bet by redesigning a Feynman diagram, which he cunningly named the "Penguin diagram". "Penguin" did eventually make it into a *Physical Review Letter* title in 1993 [39]. In a more direct manner, colleagues of mine were working on sensors to be mounted on fish. Of course they could not resist the temptation and the paper is called: *"Fish and chips: Four electrode conductivity/salinity sensor on a silicon multi-sensor chip for fisheries research"* [40].

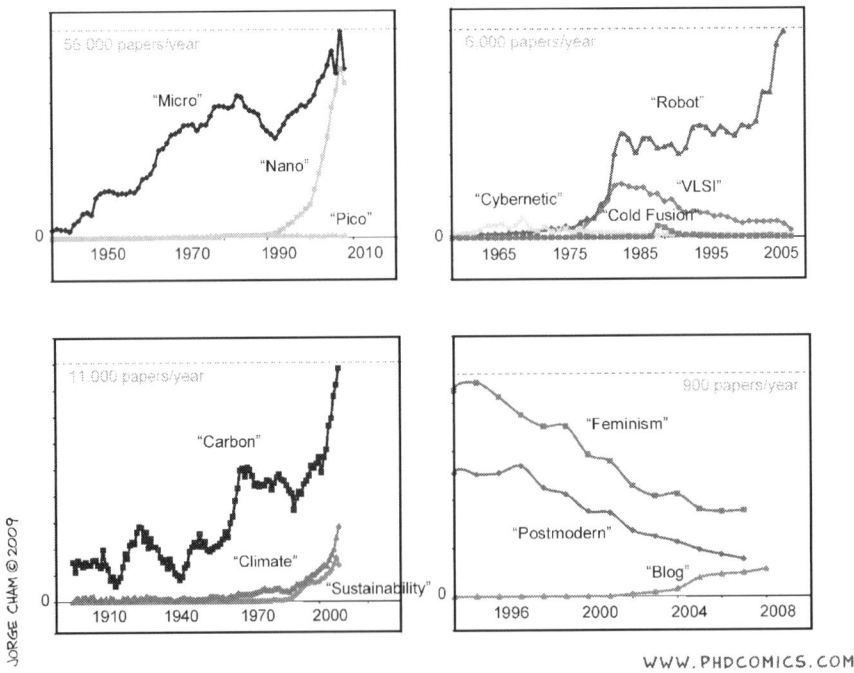

5.2. Authors

This is a very sensitive issue with two key points: Who should be authors and in what order should the authors be mentioned?

5.2.1. Who Should be Authors?

A high energy physics paper published in 2008 had 2926 authors [41]. To the outside world, it appears that if you have twiddled a screw on a particle accelerator in CERN you are entitled to be on all CERN papers forever after. It is not uncommon for PhD students in high energy physics to be co-authors on 20 papers or more by the end of their PhDs. Many high energy physicists only have self-citations because they are co-authors on all papers in the field. It makes the system somewhat difficult to judge from the outside.

The high energy physics community is a special case, but there are other examples of large-scale scientific collaborations with a very large number of authors on the papers. Papers from the human genome project, for example, can have more than 100 authors [42]. Geophysics and space science and astronomy are other fields in which papers having very many authors are not uncommon.

Large-scale scientific collaboration aside, the rule is:

- **Only people who have made an important contribution to the work should be authors, but ...**

    - It used to be and to some extent still is the case that people who were not directly involved in the work are also included as authors. It is still quite common in many places to include the head of department as author, possibly even last author (see Section 5.2.2). Several journals are now trying to counteract this by requesting an explicit statement from each author on his/her exact contribution to the paper at the submission stage (see Section 5.11.1). Nonetheless, "honorary" authorship remains a die-hard tradition.

    - If some measurements were carried out for payment, you are normally not obliged to include anybody involved with these measurements as authors. On the other hand, it might be strategic sometimes to do just this to receive goodwill and extra measurements for free some day when you are short of funding!

    - It is always good to include somebody famous in the field as co-author if at all possible. It serves as an extra quality mark.

Chapter 5. The Structure of a Scientific Paper

* Of course the names and institutions of the authors should not play a role at all, but in practice it may increase the referees' (and the editors') confidence in the work if somebody famous or somebody from a famous institution is co-author (see Section 10.1).

– **When it comes to interdisciplinary work in particular, it can be very important to deliberately extend the list of authors to make sure that it covers all relevant disciplines. This demonstrates to the referees (and eventually the readers) that all the necessary expertise is available.**

* In 2009 we wrote a technical comment to *Science* on a paper claiming a find of 30.000 year old flax fibers [43]. The methodology in the original paper was extremely poor; basically it boiled down to "*We have found these fibers. They look like flax so they must be flax*" (moral: always maintain a healthy skepticism, even with papers in the highest-ranking journals) [44]. In actual fact, all one could say on the basis of the images presented in the paper was that the fibers must be bast fibers and bast fibers are notoriously difficult to identify. When we wrote our technical comment making this point, we wanted to be absolutely sure we were making a clear statement in the community and of course also to convince the referees that we were contesting the original paper on really solid ground. The technical comment was largely written by my master student Christian Bergfjord and myself, but we made sure to include people who were acknowledged experts in all relevant disciplines on the author list: textile archeology, ancient plant DNA, archeobotany and fiber identification with X-ray diffraction.

– Generosity tends to work both ways. If you are generous, you might get the chance of co-authoring another paper in which your contribution was perhaps not so big.

– Technical staff are normally not included as authors even if they have provided major contributions. Sometimes they are mentioned in the acknowledgments. Your institute/department will probably have a policy on this.

• **Remember: If there are more than 5 authors of a paper, usually only the name of the first author will be listed when the paper is cited in other journals. The paper will be referred to as "*First author et al.*".**

People frequently differ quite substantially in their understanding of the meaning of "important contribution" and so the issue of co-authorship can lead to heated arguing and even long term animosity. Sometimes the situation is complex. What do you do for example if one student spent a lot of time building an apparatus/synthesizing a compound which is then used by the next generation of students? Clearly it is reasonable to let the student who started the work be a co-author on some of the first papers produced by new students using the apparatus/compound, but how many papers should it be? There is no simple answer to this. It will always remain a matter of judgment.

If you are a PhD student, it is important to remember that if other people approach you and insist that they want to be authors on your paper, refer them to your supervisor. Of course you can suggest co-authors to your supervisor and also say directly if you think somebody deserves to be a co-author, but at the end of the day it is a supervisor's job to decide who should be authors on a publication from the group. If your supervisor comes up with a statement like "*Oh I do not mind, you work it out between you.*" (a typical example of bad leadership), you must tell him/her directly "*Sorry, but it is your job to decide, I cannot make this decision.*". If you are first author (see Section 5.2.2), it does not matter so much how many authors are on the paper. If you think your supervisor is too generous with co-authorships in general, you can remind him/her that more than 5 authors on a paper leads to the paper being listed as "*First author et al.*" in the literature.

5.2.2. In What Order Should the Authors be Mentioned?

The order in which people are listed as authors is very important! Do not let anybody fool you into believing anything else. Some laboratories have a policy of listing the authors in alphabetical order. This is particularly popular if the head of department has a surname beginning with "Z", but it is the exception and not the rule and it is not good. The (unwritten) rules are:

- First author:
  - The person who did the main part of the work, usually a PhD student or a postdoc or occasionally a master student, should be the first author.

- Last author
  - This should be the author who was PI (primary investigator) on the project. This is the author who instigated the project and provided the daily supervision of the first-author student and typically the person who wrote the

grant application funding the project. In most cases the last author will be the group leader. This does cause a problem sometimes, for example in situations where a young assistant professor/habilitation candidate/postdoc is suffering under a head of department/group leader who thinks he/she has a God-given right to being the last author even though he/she has only had a very limited contribution to the paper. The poor assistant professor/habilitation candidate/postdoc who really needs prominent publications in which it is clear that he/she has played a major role finds him/herself squeezed in the middle. This may lead to him/her insisting on being first author and then the poor student who really should be first author is the one being squeezed.

- Middle authors
  - People who have contributed in other ways to the paper, i.e. helped take the measurements, did part of the data analysis and so on. The order here is not so strict. Students tend to be listed directly after the first author according to the importance of their contribution as far as it is possible to distinguish. Supervisors including postdocs are listed directly in front of the last author, with the most important one closest to the last author.

Notes for PhD students:

- **As a PhD student you should fight to be first author on a paper which is mainly your work.** The best way to secure that you get to be first author is to insist that you are in charge of writing the paper. You then simply put yourself as first author without asking. If you have really done a major part of the work and are now also in charge of writing, it is not very likely that anybody is going to protest.

- It is very important that you get to be first author on at least one paper during your PhD. If you come out of your PhD and only have publications where you are one of the middle authors, people may suspect that you only got included on these publications out of pity. If they see you have first-author papers as well, they are more likely to believe that you really made a contribution to your "middle-author" papers.

- If somebody else has a justified claim to the first authorship, think about the option of shared first authorship (see Section 5.2.3).

5.2. Authors

Notes for postdocs and assistant professors/habilitation candidates:

- If you are the true "prime investigator" on the project, i.e., you have played a leading role in instigating the project, written the grant application, had the scientific idea, and performed the daily supervision of the first-author student, then you should try very hard to become last author. If your group-leader/head of department does not allow this, you should think seriously about whether this person is really the right one to promote your career. Try at least to negotiate a shared last authorship (see Section 5.2.3).

THE AUTHOR LIST: GIVING CREDIT WHERE CREDIT IS DUE

The first author
Senior grad student on the project. Made the figures.

The third author
First year student who actually did the experiments, performed the analysis and wrote the whole paper. Thinks being third author is "fair".

The second-to-last author
Ambitious assistant professor or post-doc who instigated the paper.

Michaels, C., Lee, E. F., Sap, P. S., Nichols, S. T., Oliveira, L., Smith, B. S.

The second author
Grad student in the lab that has nothing to do with this project, but was included because he/she hung around the group meetings (usually for the food).

The middle authors
Author names nobody really reads. Reserved for undergrads and technical staff.

The last author
The head honcho. Hasn't even read the paper but, hey, he got the funding, and his famous name will get the paper accepted.

JORGE CHAM © 2005 — www.phdcomics.com

5.2.3. Shared First or Last Authorship: A Good Way to Increase Justice

Nowadays it is becoming increasingly popular to have shared first or last authorship. This is a very good thing, because it provides a just way of handling the fact that often it is not possible to say that one person did the main part of the work. For example, it is often the case that two PhD students work together, one doing theoretical and the other experimental work. This would be a typical situation that merits shared first authorship. In principle there is no limit to the number of authors that can be shared first authors, but more than three starts to look a bit silly.

Chapter 5. The Structure of a Scientific Paper

- For a shared first authorship paper, the first authors will be listed in alphabetical order with a footnote indicating, "*These authors contributed equally.*". If you are not the lucky one whose surname begins with "A", make sure that you make a note in the publication list of your CV that this is a shared first authorship paper.

- If two shared first-author students work in different groups, it will often be appropriate to have shared last authorship for the two group leaders.

- Shared last authorship may provide at best a fair and at least a pragmatic solution for handling difficult group leaders/heads of departments who insist on being last authors even if they have actually only had a very limited contribution to the paper.

5.2.4. Corresponding Author

The corresponding author is marked with an asterisk or a similar symbol in the published paper and the comment "*author to whom correspondence should be sent*" or something similar. Typically the email address will be listed for the corresponding author, but not for the other authors. The corresponding author is the one to whom readers can send questions about the published paper. Before the joint last authorship became a possibility (see Section 5.2.3), the assistant professor/habilitation candidate/postdoc would sometimes manage to negotiate the corresponding author's place as a "comfort prize". However, there are also cases where the group leader/head of department insists on being corresponding author on the grounds that postdocs and students leave and so for the sake of continuity in tracing work done in his/her group, he/she should do the job. In this day and age of internet, it is very easy to trace a group leader if you want to get in contact with him/her, so I do not really think this is an argument that counts, but it is often used. Personally, I handle the corresponding author question as follows:

- If a PhD student or postdoc from my group is first author on the paper, I let him/her be corresponding author. If the journal allows more than one person to be mentioned as corresponding author, I usually include myself as well.

- If a student is making a publication related to his/her master's thesis work, I do not let him/her be corresponding author. Corresponding author is a highly responsible role which, in my opinion, a person is not formally qualified to fulfill until he/she has completed a full, basic scientific training, that is, completed a

5.2. Authors

master's course of study. There is, however, no official rule preventing a master's student from being corresponding author.

5.2.5. Tracing the Work of an Author

If people have read one of your papers, they may well want to read more and they will search the literature for other papers written by you. Of course you want to make it as easy for them as possible to trace you. Here are various aspects you should consider in that context:

Author Name

- If you have a very common name, you may be difficult to find when somebody does a search on author names. Therefore, if you are embarking on publishing your first paper, think carefully about what name you want to use. If you have a very common name you might think of inventing a couple of additional initials. Instead of being simply B. Holst (of which there are several active, scientists), I could have made myself B. A. A. Holst and that would have made me uniquely identifiable. I got this idea from a Chinese colleague, who told me that it is quite common for Chinese researchers to do this. When the Chinese characters for surnames are changed into European letters, about 80% of the Chinese population share around 10 surnames.

- Some people (mainly women) may want to change their surnames because of marriage. In such cases I recommend that you get a double name. Then you can keep publishing under your old name without much confusion. Science aside, this construction also makes it easier to return to your maiden name in case of divorce.

- If you decide to change your surname completely, I recommend the approach of a male colleague. His surname was Hamburger, and I guess it was just a bit tiresome to have such a surname living in America, so he changed it, but he did so over a prolonged state of transition: In the first several publications under his new name, he wrote his old name (Hamburger) in parentheses.

- Should you use your full name rather than initials? Some journals recommend that you publish under your full name rather than under initials in order to make you easier to recognize and trace. That means I would publish as Bodil Holst, rather than B. Holst. In principle this is fine, but unfortunately several

Chapter 5. The Structure of a Scientific Paper

investigations over the years have shown that the work of female researchers, tends to be ranked lower by the referees than the work of male researchers, two fairly recent examples can be found in [45, 46]. The lower ranking of females compared to males may be due to the fact that referees (in particular for high-ranking journals) are mostly older, well-established scientists, and there are still many more of those that are male than female. We can hope and believe that this bias will disappear over the years to come, but to be on the safe side, for the time being I recommend the following:

- Women should stick to publishing under their initials, unless you have one of those strange Scandinavian Viking female names that sound like male names everywhere outside of Scandinavia i.e. Bodil, Orlaug, Torborg, Hege and Bergljot.

- Men should stick to their initials also out of solidarity, or if they have names like Andrea (male name in Italy, female name most other places) or Inge (male name in Norway, female name most other places).

- In some countries, i.e. Island and several Eastern European countries, your gender is reflected in your surname. In this case there is not so much you can do other than hope that you referee is a reasonable person, which is fortunately usually the case.

Author Identifier Code

Some publication search engines (see Section 3.8.1) provide the authors with the possibility to register themselves with a unique identifier code. This is important and something you should make sure to do. Here are two of the most widespread author identifier systems presently available, but watch out how things develop over the next years, both in your specific community and in the scientific community in general.

- *Researcher ID*; The *Web of Knowledge* has introduced the tracking system *ResearcherID*. You have to actively activate this yourself and make sure that all your publications are registered under this number. A person using the "*search for other publications by this author*" tool, will then only get a list of your publications, not all publications by authors who share signature with you.

- *ORCID*: *Open Researcher and Contributor Identifier*. Citing from the website: http://orcid.org/about: "*ORCID is an open, non-profit, community-based effort to provide a registry of unique researcher identifiers and a transparent method of linking research activities and outputs to these identifiers.*"

- *ORCID* has the additional advantage that it is connected to several journals already at the submission state. This means that when you submit your manuscript to a journal, electronically you can use your *ORCID* registration details (affiliations, email, etc.) directly as part of the submission information.
- *ORCID* and *ResearcherID* have recently linked up.

5.3. Abstract

Some journals, particularly in the medical sciences, require the abstract to be in a pre-set format, almost like a "mini-paper". In organic chemistry an abstract can be a diagram of the compound you have synthesized, but **the most common type of abstract in the scientific literature is what may be termed the "free-style abstract"**. A free-style abstract has a length limit, usually 150-300 words, but it varies from journal to journal. This section presents a structure for a free-style abstract. **You can use this structure also for writing regular conference abstracts** (see Section 3.6). In Sections 5.3.2 and 5.3.3 you will find examples of free-style abstracts using the structure presented below. Please also make sure that you read the chapter on language (Chapter 7), in particular Section 7.1 on synonyms and Section 7.4 on when to use the past and present tenses.

5.3.1. Free-Style Abstract

A good working model for a free-style abstract structure is as follows:

- **Why did we do it?** Start out by justifying your work, putting it into a broader context - why should work be done in this general area?
  - How much you want to generalize depends on how specialized the journal is, but do not leave out this important aspect.

- **What is the problem we are solving?/question we are answering?**
  - This part people tend to forget, but it is important because it places your work in the specialized context of your field and makes it easier for the reader to follow the abstract.

- **What did we do?** The main result(s) must be listed here. If the results can be quantified in terms of numbers, these numbers must be listed with error bars.

Chapter 5. The Structure of a Scientific Paper

- The take-home message (see Section 4.2) must come across clearly here! Remember, just by reading the abstract the reader should understand clearly what the take-home message is. It should in fact be possible to cite the paper after having read only the abstract.
    * One of my workshop participants once said: *"If I include the numbers in the abstract, then people will not have to read the paper"*. My answer was *"Yes that is correct, and that is a good thing!"*.
- Of course you should not include dozens of numbers in the abstract. You should just select the most important numbers that make your work appear most impressive. For example: *"The reaction yield was increased by up to 6.1 +/- 0.1 %"*.

- **What is new?** A brief comparison with the state-of-the-art (previous work in the literature).
    - People also tend to forget this, but it is vital because it helps the reader to judge how important your result is.
    - There can be cases, where the *"What is new?"* is covered by the *"What is the problem we are solving?/question we are answering"*. This is the case, e.g., if the problem we are solving can be formulated as a simple yes/no question: *"Has there been life on Mars?"*, and you are capable of coming up with a conclusive *"Yes"* answer. Remember, it is impossible to prove a negative; so the answer *"No"* is not an option (see Section 4.3).
    - Even though you refer to previous work in the literature, usually you do not include citations in the abstract (see Section 5.12). You simply write *"it has previously been shown"* or similar. The appropriate citation must then be included in the introduction (see Section 5.5). Note! some journals, for example *Nature* allow citations also in the abstract. Make sure to check this.

- **How did we do it?** Sometimes you can mention the experimental methods/theoretical tools used.
    - Very often methods are not discussed at all in the abstract. It becomes relevant if you are using a new or unusual method in your field.

A typical problem for many abstracts you find in the literature is that the first 4 points listed above simply do not come across clearly. The result is that people slightly outside the field do not understand what your work is all about.

Note: Addressing the "Why did we do it?" part before the "What did we do?" part is what one might call the "modern" style of abstract writing. An "old-fashioned" style abstract presents the "What did we do?" part first and then the "Why did we do it?" part. Alternatively, the "Why did we do it?" part is sometimes ignored completely and the abstract starts directly with "What did we do?". This is not wrong, but it is hardly the best way to reach a broad readership.

5.3.2. *Nature*'s Summary Paragraph Guideline

On the next page you can read *Nature's* official, annotated example for writing a so called summary paragraph for a letter, taken from the website http://www.nature.com/nature/authors/gta/2c_Summary_para.pdf. This corresponds to the abstract for a paper. We will now go through this summary paragraph step by step.

Chapter 5. The Structure of a Scientific Paper

Annotated example taken from *Nature* 435, 114–118 (5 May 2005).

> During cell division, mitotic spindles are assembled by microtubule-based motor proteins[1,2]. The bipolar organization of spindles is essential for proper segregation of chromosomes, and requires plus-end-directed homotetrameric motor proteins of the widely conserved kinesin-5 (BimC) family[3]. Hypotheses for bipolar spindle formation include the 'push–pull mitotic muscle' model, in which kinesin-5 and opposing motor proteins act between overlapping microtubules[2,4,5]. However, the precise roles of kinesin-5 during this process are unknown. Here we show that the vertebrate kinesin-5 Eg5 drives the sliding of microtubules depending on their relative orientation. We found in controlled *in vitro* assays that Eg5 has the remarkable capability of simultaneously moving at ~20 nm s^{-1} towards the plus-ends of each of the two microtubules it crosslinks. For anti-parallel microtubules, this results in relative sliding at ~40 nm s^{-1}, comparable to spindle pole separation rates *in vivo*[6]. Furthermore, we found that Eg5 can tether microtubule plus-ends, suggesting an additional microtubule-binding mode for Eg5. Our results demonstrate how members of the kinesin-5 family are likely to function in mitosis, pushing apart interpolar microtubules as well as recruiting microtubules into bundles that are subsequently polarized by relative sliding. We anticipate our assay to be a starting point for more sophisticated *in vitro* models of mitotic spindles. For example, the individual and combined action of multiple mitotic motors could be tested, including minus-end-directed motors opposing Eg5 motility. Furthermore, Eg5 inhibition is a major target of anti-cancer drug development, and a well-defined and quantitative assay for motor function will be relevant for such developments.

One or two sentences providing a basic introduction to the field, comprehensible to a scientist in any discipline.

Two to three sentences of more detailed background, comprehensible to scientists in related disciplines.

One sentence clearly stating the **general problem** being addressed by this particular study.

One sentence summarizing the main result (with the words "**here we show**" or their equivalent).

Two or three sentences explaining what the **main result** reveals in direct comparison to what was thought to be the case previously, or how the main result adds to previous knowledge.

One or two sentences to put the results into a more general context.

Two or three sentences to provide a **broader perspective**, readily comprehensible to a scientist in any discipline, may be included in the first paragraph if the editor considers that the accessibility of the paper is significantly enhanced by their inclusion. Under these circumstances, the length of the paragraph can be up to 300 words. (This example is 190 words without the final section, and 250 words with it.)

5.3. Abstract

The Nature Summary Paragraph annotated example fits perfectly the guideline in the previous section for how to write a free style abstract:

- **Why did we do it?** This part is covered by the first two points:
  - *One or two sentences providing a basic introduction to the field, comprehensible to a scientist in any discipline.*
    * I think *Nature* is being a bit optimistic here in the example that they use. I am a scientist, but I do not understand the first two sentences. I recommend being even more general.
  - *Two to three sentences of more detailed background, comprehensible to scientists in related disciplines.*

- **What is the problem we are solving/question we are answering?** This part is covered by the third point:
  - *One sentence clearly stating the general problem being addressed by this particular study.*
    * As mentioned in the previous section, this is a particularly important point, which is often forgotten when people write their own abstract. They remember the general introduction and present their result, but forget to state clearly the problem they are actually trying to solve. So pay special attention here and have a look at the demonstration abstracts in the next section.

- **What did we do?** This part is covered by the fourth point:
  - *One sentence summarizing the main result (with the words "here we show" or their equivalent).*
    * This is where you present the "take-home message" of your paper (see Section 4.2). A phrase along the lines of: *"here we show"*, *"here we present"*, *"this paper presents"*, *"in this paper it is shown"*, should not be omitted. People used to reading scientific papers will search for this phrase in the abstract, knowing that after this phrase the actual content of the paper (the take-home message) is presented, so use it to make life easy for the reader.

- **What is new?** (comparison to state-of-the-art/previous work): This part is covered by the last three points:

Chapter 5. The Structure of a Scientific Paper

- *Two or three sentences explaining what the main result reveals in direct comparison to what was thought to be the case previously or how the main result adds to previous knowledge.*
    * Again, this is a point that people tend to forget, but which is actually really important to emphasize.
- *One or two sentences to put the results into a more general context.*
- *Two or three sentences to provide a broader perspective, readily comprehensible to a scientist in any discipline.*
    * This last point is not always relevant, as stated.

You can see that there is very little discussion on "how" the work has actually been done. A common mistake that people make is to spend a lot of space in the abstract explaining methodology.

5.3.3. Sneezing, Cakes, Coffee: Demonstration Abstract Examples

You can practice abstract writing very well yourself, even if you do not have any results yet. In fact, in my workshops I encourage the participants to invent some amazing results and then write the abstract. This means that you can write freely without the petty restrictions provided by reality. Special thanks are due to the three workshop participants from the *Technical University of Munich* and the *University of Bergen* who gave me the permission to reproduce their abstracts with their ground-breaking (invented) results. All abstracts have less than 300 words. I have separated each abstract into four or five paragraphs, to help you to see how these abstracts relate to the free-style abstract structure described above:

- Paragraph 1: The "**Why did we do it?**" part. Here the general background and justification for the work is presented.

- Paragraph 2: The "**What is the problem we are solving/question we are answering?** or, as it is formulated in the *Nature* aid above: "*One sentence clearly stating the general problem being addressed by this particular study*". You will see that in demonstration abstracts 1 and 2 this has been extended to more than one sentence. This is fine.

    - As mentioned above, this is a very important section and it is very often forgotten. People tend just to write Section 1 and Section 3, so pay special attention here when you are writing your own abstract. Try to answer the question: *"What is it really that I want to show?"*. It strengthens your cause!

- Paragraph 3: The **"What did we do?"** part. Here the main results are presented. Here is the take-home message. As discussed above it is a very good idea always to start this part of the abstract with a sentence like "*Here we present*" or "*Here we show*" or similar to make it clear to the reader that now he/she needs to pay attention, because now they are going to learn what has been done in this piece of work.

- Paragraph 4: The **"What is new?"** part. This is important to highlight what is new in your work, why your "take-home message" is particularly important and interesting.

- Paragraph 5: The **How did we do it?** part. This is the methodology section, only included in abstract 3, where the methodology is particularly important for the results.

Abstract 1 - A Device for Extracting the Mechanical Energy of Sneezing

A Device for Extracting the Mechanical Energy of Sneezing

K. Müller

Technical University of Munich

1) The increasing use of mobile, electronic applications such as mobile phones, ipads, etc. has increased the need for transportable energy devices. Usually rechargeable batteries have been used; however, the regular recharging can be inconvenient. This together with environmental issues makes it desirable to use renewable energy devices, which can be recharged on-site. At present only a few such devices exist, (i.e. solar cells), and they tend to be expensive and/or limited in efficiency. Hence there is a need for new approaches.

2) During the spontaneous rapid air expulsion (popularly referred to as sneezing) of a healthy, average-sized adult (70 kg), the air is accelerated to more than 200 m/s. This phenomenon has long been considered a promising source of renewable energy. However, a versatile, efficient method of energy conversion was lacking.

3) Here we present a solution to this pressing matter. We have developed a micro-scale, turbine-shaped generator with an efficiency of 0.41. When plugged onto the nose, this generator is capable of delivering up to 200 W from a single sneeze. A capacitor attached to the generator stores the electrical energy generated. With a single sneeze, we can light two 100 W light bulbs for 0.32 ± 0.04 s.

4) The generator can deliver at least a factor 2 more power than other renewable energy devices described in the literature. A person suffering from a moderate cold, corresponding to a sneezing frequency (f) $\leq 0.2\ s^{-1}$ can easily deliver the electrical energy to power e.g. an mp3 player or the LED-headlight of a car. The more severe forms of a cold or an equivalent pollen allergy (f $\geq 1\ s^{-1}$) can power even larger electronic applications such as e-bikes. The generator can thus have a considerable impact also on e-mobility.

This is in principle a very easy abstract to write, because there is one clear, good result. Note how all the crucial numbers are presented in the abstract: Energy efficiency, energy production from a single sneeze and two quantified examples of what can be achieved in realistic scenarios (moderate cold and pollen allergy). This is really an abstract you can cite without reading the paper (see Section 4.2).

The very last line states that the new generator can have a considerable impact on e-mobility. This should not be mistaken for an outlook on future work, which should not be in the abstract, but only in the conclusion. Rather this remark puts the result in a context. This is a bit of a gray area, but in the abstract you should only mention things that can be done now, without further research and development. The idea is that the generator is already so good that it can have a considerable impact on e-mobility now, without further improvements. If you want to make it really perfect, you could try to quantify the impact, but as a final remark you can get away with a slightly "hand waving" expression like "*considerable*".

Note further how the words "*applications*" and "*devices*" have been used for separate issues. In the introductory sentence, one might as well have talked about "*mobile, electronic devices*"; it may have been the most natural choice in fact, but this would have created a potential confusion with the reference to "*energy devices*". By avoiding the use of the word "*devices*" in two different contexts, the abstract is more clear. The generator could also have been referred to as a "*device*", but the word "*generator*" has been used consistently throughout the abstract. The importance of always using the same word for a particular thing or concept is described in detail in Section 7.1.

Abstract 2 - The Quality of Sweet Pastry can be Determined by Fruit Fly Spotting

The Quality of Sweet Pastry can be Determined by Fruit Fly Spotting

M. Blume

Technical University of Munich,

1) Numerous investigations have shown that sweet pastry is crucial for the well being of mankind. Hence, in order to increase the general well being it is important to be able to determine the quality of sweet pastry in an objective manner.

2) Fruit flies have long been known as good indicators for the quality of sweet pastry. However, up till now, no standardized quality assessment method has been established. Different approaches have been applied, leading to conflicting results in the existing literature and making a comparison between different types of sweet pastry difficult.

3) Here we present a new method capable of assessing the quality of sweet pastry. The new method is based on deterministic fruit fly spotting. A crucial novelty is the introduction of two-dimensional boundary conditions, determined by the sweet pastry surface shape. The new method was developed using experimental data from more than 20 different types of sweet pastry including Danish pastry, fruit pies, treacle tarts, brownies and cupcakes.

4) With this method, it is possible for the first time to assign a unique quality-parameter (which we label the fruit fly number F) to sweet pastry, independent of pastry type. The experiments revealed that there is a maximum packing density of fruit flies on the surface of sweet pastry. Further it was shown that fruit flies follow a deterministic "all-on-one-target" route. In other words, rather than spreading evenly over all pieces of pastry, the fruit flies initially prefer the same piece of pastry and only start spreading to other pieces when the maximum packing density has been reached. These two experimental findings offer an explanation for several conflicting results in the literature. They have both been included in the new method presented here.

This abstract illustrates and successfully handles a typical problem in an abstract. We have one key message - The new method, which allows different types of pastry to be compared. But we also have the two new experimental findings which have been

Chapter 5. The Structure of a Scientific Paper

included and are important for the method, but which are also important in their own right, because they can explain previous conflicting results in the literature. The way to handle this is by tailoring the abstract so that the emphasis is on the key message: the new method. The conflicting results in the literature are mentioned only as a second point in paragraph 2 and 4 and the new method is mentioned again at the very end of the abstract to emphasize its importance. Note also, how the title is a statement: The main take-home message of the paper (see Section 5.1.1).

Abstract 3 - How Much Coffee? A Novel Method for Determining and Optimizing a Person's Concentration Level

How Much Coffee? A Novel Method for Determining and Optimizing a Person's Concentration Level.

A. Kristoffersen

University of Bergen

1) A concentrated mind is important for intellectual work. Hence it is important to determine how to stimulate and optimize the concentration level.

2) Several approaches have been suggested, but up till now, no general solution has been available.

3) Here we present a novel brain wave scanner, which can be used under natural working conditions. We tested the scanner by analyzing the brain wave pattern for 200 test persons (100 male and 100 female) drinking coffee, a well known concentration stimulant. The combination of different types of brain waves indicates over- or under-stimulation, and thus a person can adjust the intake of coffee accordingly. We found that the peak concentration level for a test person (male or female) occurs at 3.8 ± 0.2 cups of coffee in a period of two hours.

4) Previous results suggest 2 cups of coffee within two hours as the optimum. However, these results were obtained using standard scanners that did not allow natural working conditions.

5) Our novel brain wave scanner is based on an extremely sensitive voltmeter (pico-volt range). It measures the electrical activity of the brain and is equipped with unique software capable of analyzing the brain wave pattern and deducing from this how receptive and efficient the brain is in real-time. Combined with a noise-eliminating algorithm, this makes it possible to record the electric potentials generated by neurons in the brain, as demonstrated by the authors in a recent Nature article. The algorithm used to determine the relationship between brain wave pattern and concentration level has been developed with empirical data obtained by exposing the 200 test persons to mental tasks and evaluating their performance in combination with measuring their brain wave patterns.

This abstract includes a paragraph 5 on how the work has been done (methodology). As discussed above this is not always appropriate, but for this particular abstract it is sensible, because the description of how the new brain scanner works is of particular importance. Note also the specific reference to recent work by the author published in *Nature*. From a strictly scientific point of view it would have been enough to write "*as recently demonstrated*" instead of "*as demonstrated by the authors in a recent article in Nature*" and some people might say that it is too much to include this specific journal reference. My slightly cynical approach is that it will serve to immediately impresses the people who read it (most importantly the editors and referees) and they will read the rest of the paper in a very positive mind frame. Even if the reference had been to work in *Nature* not done by the authors themselves, I would still recommend including it: "*as demonstrated in a recent article in Nature*". This helps to reinforce the impression that this is a paper on a really important, "hot" topic, which any journal would be glad to publish. It is quite possible that the referee or the editor will eventually request a reformulation, removing "*Nature*" from the abstract. This is not a problem, it would have served its purpose helping to get the paper published. The topic of how to handle the referee comments is treated in Chapter 14.

5.4. Key Words and/or Subject Classification Numbers

In many papers there is a small section, just one or two text lines between the abstract and the introduction. Here a series of key words or subject classification numbers are listed. The aim of these key words or numbers is to make it easier for a reader to search for the relevant papers. To some extent, key words and identification numbers can be seen as a left-over from the days where the publication search engines (see Section 3.8.1) were not so powerful. Nowadays it is not a problem to conduct a search for key words in the whole paper.

- If key words are requested by the journal where you want to submit your paper, you should include them even if you think them superfluous, because the more

Chapter 5. The Structure of a Scientific Paper

you comply with the journal guidelines, the better. It is important that your key words are consistent with the key words used in other relevant papers. A strategy for selecting the best key words could be to look at the key words for the most important papers you are citing.

- Some publishers operate with subject classification numbers. The so-called PACS numbers: Physics and Astronomy Classification Scheme, were introduced by *The American Institute of Physics* (AIP) in 1970 to make it easier to search for scientific information. Reflecting the fact that publication search engines are now so much more powerful than back then, AIP decided to discontinue PACS in 2010, but some journals still use it. AIP state on their website that they are working om a new classification scheme that will be made available in the near future.

If you do interdisciplinary work, you should be very aware of the fact that the same word can mean different things in different scientific disciplines. For example the word "substrate" in surface science means the surface on which the process you are investigating takes place, where as "substrate" in biochemistry refers to the molecule acted on by an enzyme to produce a product.

5.5. Introduction

The introduction is an important part of the paper and the time needed to write a good introduction should not be underestimated. **The purpose of the introduction is to place the work you have done in a scientific context.** This is done by addressing the points below, in the order they are listed.

- A clear motivation for the work (why are we doing it) with an outlook on the field and why it is important in general.
    - Start your "field definition" broadly and narrow down, step by step, to your particular subfield. I.e. surface science ▶ insulating surfaces ▶ techniques for investigating insulating surfaces ▶ molecular beams ▶ neutral helium microscopy.
        * Make sure that each field step is clearly identified using key words (often words from your title). This applies in particular to your final subfield.

5.5. Introduction

* Limit the number of steps when you narrow down. More than 5 steps are too many. Fewer are fine.

  · The more you narrow down, the more you write: just one sentence on surface science in general, two sentences on insulating surfaces, one paragraph on techniques for investigating insulating surfaces, and so on.

  · You should include citations at each step. If there are relevant review papers, this is the place to cite them.

– The danger is that you might discuss information that is too general, which leads to a loss of focus.

– The other danger is that you might narrow down your field too much too quickly thus losing the interest of the reader.

  * This can be a difficult balance to strike!

- A clear formulation of the purpose of the work: What is the problem we are studying?

- A summary of the current understanding of this problem must be given through a clear overview of state of the art/work previously done in the field with citations.

Chapter 5. The Structure of a Scientific Paper

- The key point here is to summarize what was known before you did your work. Concentrate on facts and the final outcome of previous work. For example, you should not explain how previous experiments were performed. If the reader wants to know more, he/she can read the original papers that you cite.

    - If there are conflicting or contradicting results in the literature, state this, but do not provide explanations. This is a topic for the discussion.

- A brief explanation on why you approached the problem the way you did and what the outcome was. That is to say the major result(s) should be mentioned, but just in a general manner emphasizing why the result is important and what it reveals.

    - Detailed explanations of methods should not be included here. That belongs to the methods section, but if you are using a novel method in the field, then it is appropriate to mention it in the introduction.

    - Of course you should present your work in the best possible light, but this is not done by writing very general, boasting sentences, like "*the work presented here is ground breaking*". There are some journals who even forbid the use of words like "*new*" or "*novel*" in title and abstract.

    - **Never write anything negative about other people's work**. If for example you are using a new method in your work, do not write: "*the new method is much better than previously applied methods because..*". This is the kind of things that can irritate referees (who might for example have worked with one of the other methods for the last 20 years). It is also not fair, because there is never such a thing as a "better" method. An electron microscope is not "better" than an optical microscope just because it has a higher resolution. For a start, it is usually a lot more expensive and requires vacuum conditions. The correct way of phrasing the sentence above would be *"the new method is particularly suitable for the experiments presented here because ..."*

The length of the introduction will vary depending on the type of paper you are writing (see Section 3.1). Sometimes it will just be half a page with a couple of paragraphs, and sometimes a separate section with its own header, two-three pages long. Regardless of the length **the introduction must always contain the following information: What is the question we are answering/problem we are studying? Why is it an important question/problem? What did we know about it before this study (state-of-the-art)? How do the results presented advance our knowledge?**

5.5.1. Motivation for the Work

The first part of the introduction can be seen as an extended version of the first part of the abstract - the "*Why?*" part. You should always start the introduction by describing the general area that you are studying and why it is important. Why did we do this work? Why is it interesting? Why now? Have there been any developments recently, scientifically or otherwise, which make it particularly timely to have done the work now?

- Do not worry about repeating sentences from the abstract in the introduction.

- If the journal does not allow citations in the abstract, make sure that citations for all the work you refer to in the abstract are included in the introduction.

- Make sure also to repeat the general "problem to be solved" here very clearly - you can elaborate more than you did in the abstract.

5.5.2. State-of-the-Art/Previous Work Done in the Field

The second part of the introduction should consist of a careful description of work done in the field so far (the "state-of-the-art"). Some journals require a separate "state-of-the-art" section after the introduction, but if this is not the case, then it should be covered by the second part of the introduction.

- The state-of-the-art section of your paper is arguably the most important section after the results section. This is where you show that you have done your scientific homework in the form of background reading and general knowledge of the field as demonstrated by an extensive list of all relevant citations.

- Very careful and detailed literature research is absolutely crucial for writing a good state-of-the-art section. It is not good to forget important papers.

  – Literature research is not easy. An aid to make sure that you have not forgotten any important citations is to take the papers most relevant for the manuscript you are working on and do a "forward search" on these papers in the publication search engine that you are using (see Section 3.8.1). A forward search looks for all papers citing a particular paper: If a paper cites a paper that was relevant for you, the citing paper may be relevant as well.

  – Another approach is to search for papers that have two or more references in common with papers you know are particularly relevant for your work. This option is available in the *Web of Science*, for example.

  – Search for new papers by authors who have already contributed to the field.

  – Normally you will not cite textbooks, encyclopedias or similar sources, because the knowledge here can be considered general knowledge within the discipline.

  – It is always very good to cite relevant review papers.

  – If you find out after your paper has been published that you have forgotten an important reference, it is perfectly possible to write to the corresponding author (see Section 5.2.4) of that particular paper and apologize or alternatively approach him/her at a conference. I have done this on a couple of occasions and my experience is that it is a good thing to do. I have had comments like *"I did wonder, but now I understand; this is the kind of thing that can happen, that is fine..."*. Basically, if you do not apologize, people may think you did it on purpose and may be annoyed.

    * People are sensitive about not being cited because citations are important not only generally (see Section 2.3) but also more specifically for establishing your reputation in a field. A colleague of mine did the first experiment in an area that developed into quite an important line of research, which he did not pursue further himself for various

5.5. Introduction

reasons. His first experiment was followed up by a second experiment done by a very famous group at a very famous American university. In their first paper they do cite my colleague, but in all papers they have published since then, they have only cited their own paper followed by the comment "*and references therein*". In this way they cite the paper of my colleague without citing him and since he is no longer active in the field, he has no way of "citing back".

- A sloppy state-of-the-art section will NOT impress the referees. The tendency will be to infer: sloppy state-of-the-art = sloppy science.

- **When you discuss previous work, it is important that you do this in a strictly factual manner. Remember: Never write anything negative about other people's work and do not write anything that could be taken as a criticism. Be very careful about your formulations. Sometimes it can come down to seemingly minor nuances that determine how your sentence comes across.** The point is that impolite statements might upset the reader and even worse the referee:

    - *DON'T write: Previous work did not consider B*
        * *DO write: Previous work focused on the exploration of A and did not include B*
    - *DON'T write: The approach of XX is wrong because ...*
        * *DO write: The approach of XX assumes C, which is not applicable for the experiment presented here because ..*
    - *DON'T write: The resolution is poor*
        * *DO write: The resolution is limited*
    - *DON'T write: Previous results are wrong because ...*
        * *DO write : Previous results differ from the results presented here.*

Chapter 5. The Structure of a Scientific Paper

5.5.3. Paper Outline

If you want to, you can finish the introduction with a short description of the paper outline. If you are short of space, the "paper outline" part of the paper can be omitted. Some people think it should be omitted in any case because it is redundant information (repetition). My opinion is that space permitting, it is a nice way to round off the introduction and remind the reader once again about what the "take-home message" of the paper is. In general I am a fan of a modest amount of redundant information because it helps the reader to find out if he/she has really understood what the paper is about. The paper outline part is short and can essentially be boiled down to a paragraph reading roughly as follows (see Section 5.13 for an explanation of "online supplementary material"):

- *In this paper we show that "take-home message". A description of the experimental setup used to obtain the results can be found in Section 2. Section 3 presents the theoretical model used for analyzing the results. The actual results in comparison to the theoretical model are presented in Section 4. The paper finishes with a discussion/conclusion in Section 5, followed by the Acknowledgments in Section 6. Detailed information about the experimental methodology can be found in the online supplementary material document.*

5.6. Methods/Experimental Setup

The original purpose of a scientific paper was to enable other people to reproduce the experiment presented in the paper, in case it was not possible for them to attend a demonstration session (see Section 1.1). According to this way of thinking, **the methods/experimental setup section should ideally provide a "cooking recipe" for obtaining the results presented in the paper. Further it should be convincingly argued that the results presented are trustworthy. This is done, i.e. by discussing sources of errors and repeatability. The methods/experimental setup section ought to be the most detailed part of a paper.** It follows that the only justification for making the methods/experimental setup section quite brief would be that the results presented were based on a method described in detail in a separate paper. Sometimes this can actually be a very nice way to present your work, (see Section 3.1).

Modern reality is a bit different. Some journals only allow a brief methods section to save space. Furthermore, it is usually not possible for a modern referee to check if the methods section is really correct. This would require him/her to actually reproduce the results in the paper on the basis of the information in the methods section and this is mostly not possible within a reasonable time scale. Hence it is easy for authors to get away with badly written/incomplete methods sections and indeed they often do. If you find that you cannot repeat an experiment or a calculation on the basis of the information presented in the methods section of a paper, do not despair. Most likely it is not you who are being stupid, it is the method section which was poorly written (perhaps deliberately) to make it harder for the competition.

Chapter 5. The Structure of a Scientific Paper

YOUR IRREPLACEABILITY

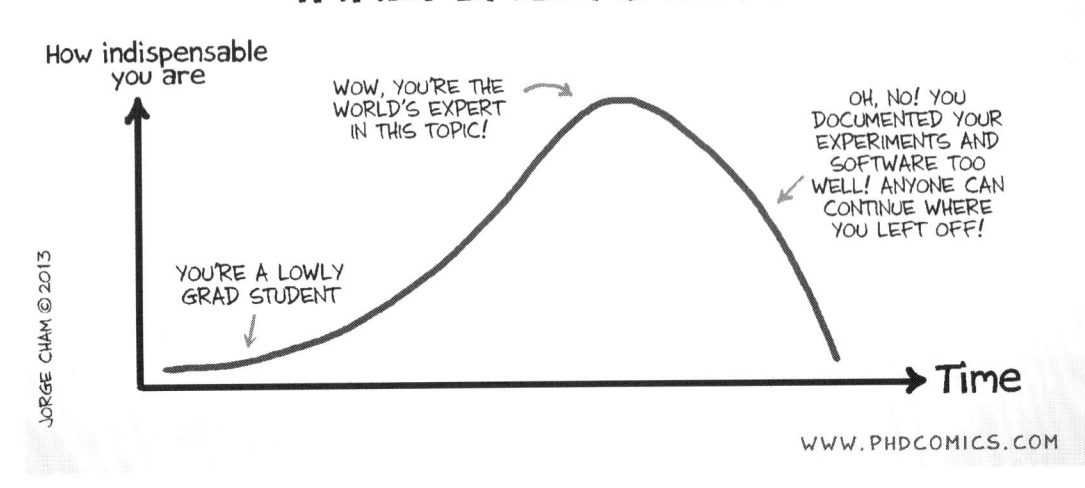

How indispensable you are

YOU'RE A LOWLY GRAD STUDENT

WOW, YOU'RE THE WORLD'S EXPERT IN THIS TOPIC!

OH, NO! YOU DOCUMENTED YOUR EXPERIMENTS AND SOFTWARE TOO WELL! ANYONE CAN CONTINUE WHERE YOU LEFT OFF!

Time

JORGE CHAM © 2013

WWW.PHDCOMICS.COM

- Let me emphasize this once more. It is not uncommon that a methods section is deliberately poorly written to prevent competing groups from doing further work in the area. It seems to be particularly common in the life sciences where everybody has more or less access to the same methods. Let's face it, it is quite safe for somebody like me to write a very careful methods section because I work mainly with a home-built ultra high vacuum apparatus, which weighs 1.5 tons and was originally constructed at the Max Planck Institute for Fluid Dynamics in Germany. (If you get the chance to be a Max Planck Director, say yes!). To do my type of experiments, you would have to build a similar instrument, which would cost you several hundred thousand euro and require access to a very skilled mechanical workshop.

    – In a deliberately poorly written methods section, nothing directly wrong will usually be written, but crucial steps may be missing or described too vaguely.

    – If you have the suspicion that the methods section for a paper has deliberately been written in an unclear manner, you can try to obtain more information in the following ways:

        * If the main author is a PhD student who has already graduated, you

5.6. Methods/Experimental Setup

may be lucky and find a more detailed description in his/hers PhD thesis. The thesis will be available through the library at the university from which the student graduated.

* Approach the main author or the group leader directly and ask. The danger here is that you may have to tell them what you plan to do and you may not want to do that.

* Best way: Get hold of the main author at a conference and invite him/her out for a couple of drinks.

Oh no, your paper exceeds the maximum number of pages allowed! What do you do??

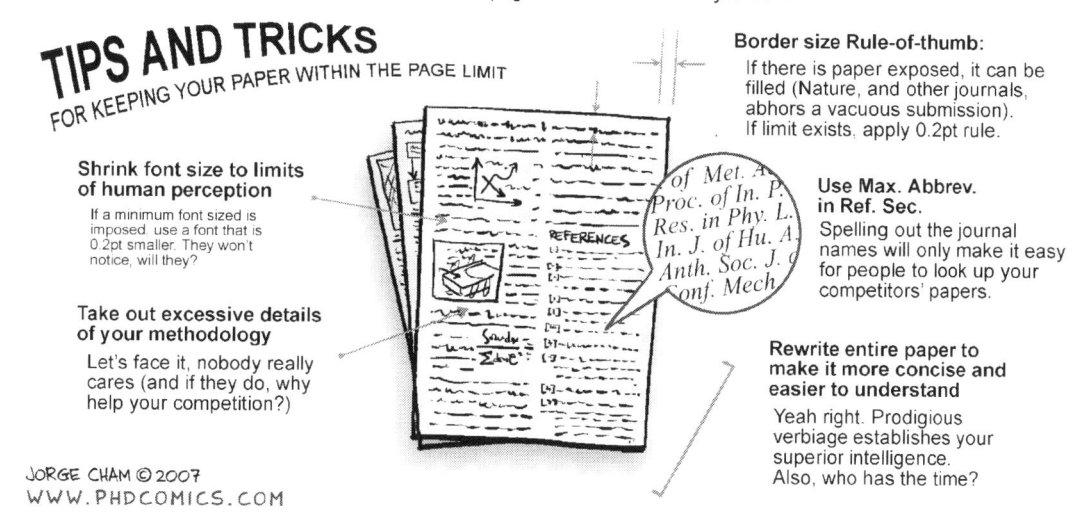

TIPS AND TRICKS FOR KEEPING YOUR PAPER WITHIN THE PAGE LIMIT

Shrink font size to limits of human perception
If a minimum font sized is imposed, use a font that is 0.2pt smaller. They won't notice, will they?

Take out excessive details of your methodology
Let's face it, nobody really cares (and if they do, why help your competition?)

Border size Rule-of-thumb:
If there is paper exposed, it can be filled (Nature, and other journals, abhors a vacuous submission). If limit exists, apply 0.2pt rule.

Use Max. Abbrev. in Ref. Sec.
Spelling out the journal names will only make it easy for people to look up your competitors' papers.

Rewrite entire paper to make it more concise and easier to understand
Yeah right. Prodigious verbiage establishes your superior intelligence. Also, who has the time?

JORGE CHAM © 2007
WWW.PHDCOMICS.COM

The content of the methods/experimental setup section of a paper is of course very field dependent and hence difficult to discuss in detail in a general book like this. Fortunately it is usually an easy section to write. If you are writing your first paper, you can very often seek inspiration in previous papers from your group, just make sure that you do not copy the text directly. If you have already written papers in the field you can do quite a bit of recycling (see Section 6.1.1).

Though this is not the way it should be, the reality is that a quite poorly written methods section is rarely a reason for rejecting a paper.

If you are a conscientious scientist and want to write a good methods section to ensure maximum progress for humankind, you must consider the following points:

Chapter 5. The Structure of a Scientific Paper

- **Make sure to list all information about commercial experimental equipment/software that you use (name, producer, etc.) even if you think it is completely standard and not necessary**. Remember, people may use the results in your paper in a completely different way from what you originally imagined, and then they may need information you did not think was important. No paper was ever rejected for providing too detailed information. A possibility, in particular in the case of space limitation, can be to include some of this information in a separate "online supplementary material" document (see Section 5.13).

- For experimental work: Provide a detailed protocol (cooking recipe) for how the data was obtained. Discuss issues such as: controls, treatments, stability tests, how many samples were collected. Make sure that you justify and explain your methodology point for point, as a clear way of arguing that your results are trustworthy.

- It is usually appropriate to include a diagram of your experimental setup.

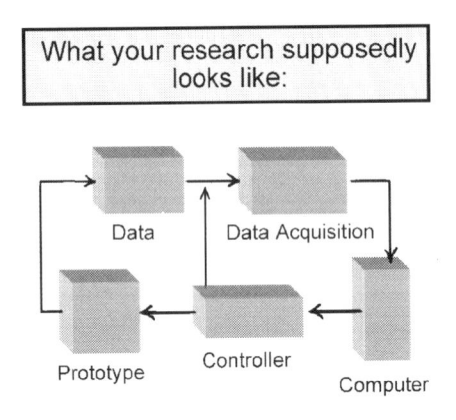

Figure 1. Experimental Diagram

Figure 2. Experimental Mess

- A description of the statistical data analysis is in principle something which should be included in the methods/experimental setup section, because this is related to the quality of the data. This is different from the analysis of the results,

5.6. Methods/Experimental Setup

which is related to the fitting of the measurements to a model, for example. This should be discussed in the results and analysis section.

- It can sometimes be very useful for the reader if you provide a brief discussion about unsuccessful attempts you may have undertaken in order to reach the results presented. Perhaps you can describe even unsuccessful attempts at reaching beyond the results presented - provided you have given up trying, otherwise it might be helping the competition a bit too much. Because it is not possible to publish negative results (see Section 4.3), the only place you can sneak in such potentially very useful information about unsuccessful approaches is in the methods section of a paper.

Chapter 5. The Structure of a Scientific Paper

THE METHODOLOGY SECTION TRANSLATOR

What it says:

"All procedures were approved by the Internal Ethics Review Board"

"Samples were treated with 0.03% sodium citrate buffer for 60.3 min. at 37.4 deg with 20.5 mg/kg poly(I:C) dissolved in 0.97% sterile PBS volume of 8.2 ml/kg"

"The solution was isolated using catalyst CH2Cl2/Et2O 4:1 in 71% yield as a mixture of 1 H NMR (CDCl3) δ 7.90 (ddd, J = 3.2, 5.2, 20.4 Hz, 1H), 7.30 (dd, J = 0.8, 2.0 Hz, 1H)"

"Measurements were performed with $-1.74 < \eta < 1.74$ around a field of 1.16T with $\sigma(pT)/pT \approx 0.5\%\ pT\ /GeV + 1.5\%$"

"Experimental kits from a commercial vendor were used and applied according to the manufacturer's instructions."

"Filter and gain settings varied with experimental conditions and objectives."

"Simulation parameters were chosen based on empirically realistic values."

"The treated preparation was incubated overnight."

"Analysis was performed using a commercially available software package."

"Statistical significance was assessed using the Student's T Test."

What it really means:

"Please don't come protest outside our lab."

"If you deviate from this by one number, it's not my fault when you can't replicate my results."

"My advisor has no idea what this means."

"I don't know why this works but this is how the previous grad student taught me to do it."

"We wasted a lot of time trying to do it ourselves, but it turned out you can just buy it."

"We twiddled the knobs until it worked."

"We made stuff up."

"I went to have a few beers with my friends."

"I put the numbers into this magic box and out came my thesis!"

"Yes, all that just to verify it with something they teach in High School now."

JORGE CHAM © 2012   WWW.PHDCOMICS.COM

5.7. Theoretical Background

Many papers do not have a separate section on theoretical background. A separate theoretical background section is relevant for a paper in two cases:

- If you are doing a theoretical paper and want to present the background for your own work with some technical details because you need these technical details to describe your own work in the results section. Such technical details would not be appropriate to include in the introduction. In the introduction you should only describe in words the previous work done.

- For a paper based on experimental work, the theoretical background section is used to present the model that you use to analyze (fit) your experimental results. The analysis (fitting) itself is described in the results and analysis section.

The following points are particularly important to address in this section:

- At the beginning of the section, make sure that you list all the basic assumptions/restrictions that are used, i.e. constant temperature, no pressure increases beyond a sudden limit, etc.. Even if you think some of these restrictions are obvious and well known, include them anyway. This is a great help for newcomers in the field.

- Make sure that you are consistent in your use of labels and symbols. Use the same labels and symbols throughout the paper. Further, make sure that all labels and symbols are clearly explained. Check with previous articles if there are some symbols that are standard for particular applications. It is very confusing if you use labels differing from the standard, i.e. imagine a paper where "T" was used as a symbol for pressure and "p" was used as a symbol for temperature. Formally there is nothing wrong with that but it would be very confusing.

- It is OK to present new work in the theoretical background section. This could be the case if you extend a model beyond what has previously been published in the literature in order to fit your experimental conditions. It is important that you make it clear what is new and what is already established in the literature.

Chapter 5. *The Structure of a Scientific Paper*

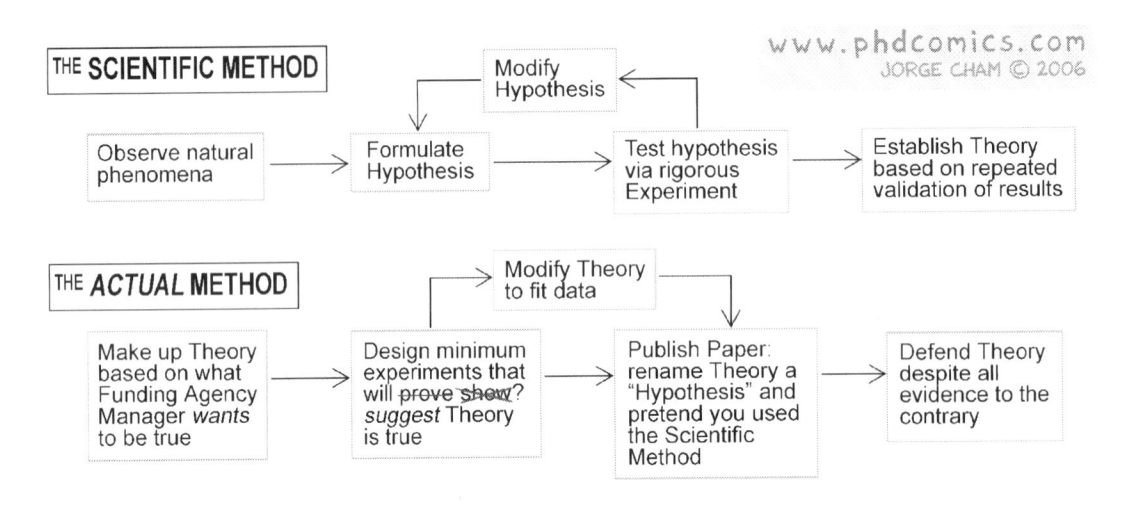

5.8. Results and Analysis

This is the section where your data/calculations are presented, sometimes in comparison to theoretical models (see Section 5.7). This is typically done using tables and figures, unless you are writing a very theoretical paper, in which your result is a mathematical derivation. For this reason and in most cases, the crucial part of writing the results and analysis section is actually to prepare the figures. Therefore: **Regardless of how you structure your writing process, you should have the figures and tables presenting your results ready, at least in draft form, before you start writing the results and analysis section.**

- A good way to start the results and analysis section can be simply to write: *"Our main results are presented in Figure 2. As can be seen, there is good agreement with the theoretical prediction described in the previous section, etc.."*
    - It may not be Figure 2 of course, but it can never be Figure 1 since you should always have at least one figure in your methods/experimental section session with a diagram of your experimental setup.
- All symbols used in the figures must be explained in the figure caption. It may seem obvious to you, but it is not always obvious to others.
- Clearly indicate each data point on the graph. The best way to present data points is using circles.

5.8. Results and Analysis

- If you need to compare two data sets, choose filled and empty circles.

- Make sure that the figure legends and scales and data points are large enough so that they are clearly visible.

  - Be aware! This means large enough when printed in the journal format. If you write in *LaTex* you have the nice option of seeing your paper in the style of the final printed version, so then it is easy to check. If you do not have this option, make sure to print the figure in a reduced version roughly corresponding to the final size.

- Always show error bars. If the error bars are so small that they do not show up on the scale, then explicitly state that in the figure caption.

- **Think very carefully about what message you want to convey in the figure. The figure is your tool to convince the reader that your "take-home message" is correct.**

  - Remember how people read papers (see Section 4.2). First you read the abstract, then you look at the figures. The take-home message must be understandable at all levels, so **use the figure caption to guide the reader in how they look at the figure. Tell them in the figure caption what they are supposed to see in the figure.** Do not just leave explanation to the main text. Write sentences in the figure caption that do just this: *"This peak shows.."*, *"The drop in the curve indicates.."*.

    * I was inspired to emphasize the point above during one of my paper writing workshops. One of the participants was preparing a paper on image-treating algorithms. In the paper draft there was a figure showing two seemingly identical pictures of a street with cars and trees and so on. The figure caption read: *"To the left an image is shown treated with standard algorithm X, to the right an image treated with the new algorithm presented in this paper."*. I told the workshop participant: *"These two images look exactly the same."*. The answer I got was: *"Yes isn't that great?"*. It turned out that the whole point was that the new algorithm did not cause any visible image distortions. We agreed on an additional sentence in the figure caption: *"The two images look identical, which illustrates the power of the new algorithm."* Of course this was quantified properly in the main text.

5.9. Discussion

The discussion section is often included as part of the "results and analysis" or "conclusion" section. Personally I prefer to include the discussion at the end of the results and analysis section if the journal allows it, because I think the conclusion section reads best if it is quite short. The purpose of the discussion section is, as the name suggests, to discuss the results. Usually there are only two or three major points to discuss:

- How well do the results agree with each other?
  - This is relevant if you measure the same parameter with different methods.
  - A short discussion on repeatability/experimental precision might be relevant, in particular if improvements are needed for further conclusions to be drawn.
- How well do the results agree with previous work in the literature?
- How well do the results agree with the theory?
  - Typically you will have a theoretical model that you use to fit your data and then you discuss how good the agreement is. Remember if you have a disagreement, it is not enough to simply provide a description along the

lines of *"here the results agree, here they deviate"*. The reader is capable of seeing for him/herself. You have to provide some suggestions for why your agreement is not so good. This can be quite "hand waving" but should not be omitted. Write a sentence like: *"This deviation results from/is due to/arises from/is caused by ..."*. If you are not so sure about the reason you can write: *"This deviation is most likely due to..."* or *"This deviation most likely results from..."*

- Alternatively, you are testing a hypothesis, for example: *"Does this drug have an effect?"*. You then discuss your results on the basis of this hypothesis. To what extent do the results confirm the hypothesis (This should be clear from the statistical analysis of your results.). Are there any unexpected effects?

5.10. Conclusion

The conclusion is more or less a repetition of the abstract with the exception that there is no "Why?" bit. Instead a discussion of "future work" is included. In former times when there were fewer people working in science and communication between people in general was much more limited, the "future work" was very important. It served to inform your colleagues what you were working on so that nobody else would start doing it. A sort of gentleman's agreement. The world does not work like that any longer.

Because there is such a large overlap between abstract and conclusion, some journals tell you to omit the conclusion section completely and write an extended abstract which includes discussion of future work instead. However, for many journals a conclusion section is still standard. You can write it in the following way:

- First you list the major result achieved: *"In this paper we show/In this paper we have shown..."*. You can elaborate a bit more than in the abstract, but do not include dozens of numbers.

- Then you summarize the open questions raised through the results presented in your paper. For example:

  - *"The agreement with theory is not so good above the temperature T."*

Chapter 5. The Structure of a Scientific Paper

- Finally you describe future work to do.
  - The easy way out is to sketch roughly how one could hope to answer some of the open questions. For example:
    * *"More experiments should be performed above the temperature T to clarify the behavior. This will require a modification of the present experimental setup."*
  - Note: Only list obvious future work. If you have some clever ideas, do NOT list them. Science can be very competitive and people sometimes play dirty tricks, so be careful.
  - Do not make the future work tasks sound too simple, because then the reader may wonder why you haven't done them already.
    * This is why the remark about the necessity to make a modification of the experimental setup before more temperature measurements can be made has been included above.

5.10. Conclusion

5.11. Acknowledgments

The acknowledgments section is optional, but in practice it is highly recommended to acknowledge your funding agency. This is becoming increasingly important, to the extent that **some funding agencies (the European Commission for example) do not allow you to list a paper in your final project report unless you have acknowledged the funding agency in the paper.**

Apart from the funding agency, you should thank people who have contributed to the project in some way but not so much that it merits a co-authorship. For a discussion on who should be author on a paper, see Section 5.2.1. Typical "acknowledgment contributions" are discussions, proofreading the manuscript, technical staff who might have constructed part of the instrument or performed parts of the experiments and referees if you found their comments particularly helpful. You list the names of the people, with a brief description of what you thank them for. You do normally not list any titles or affiliations. The referees you just refer to as referees.

You should always ask for permission before you acknowledge somebody apart from the referees, because they may be so modest that they do not think that their contribution merits an acknowledgment. This may just be a polite way to express that they do not want to be associated with your work of course, but modest people do exist even in academia.

Be aware that if you include somebody in the acknowledgments that person is not likely to be picked as a referee by the journal.

Be sure that you do not forget anybody. Some people get hurt if they do not get an acknowledgment even though it does not matter at all from a prestige point of view (There is no space in an academic CV where you list the papers in which you have been acknowledged!).

The most delicate situation is when you approach somebody and they think that they should be co-authors, not just acknowledged. There is no easy solution to that problem, but at least it is better to have the discussion up front rather than create an enemy for life when the paper is published (see Section 5.2.1).

Acknowledgments

What you write:	What you actually want to say:
"Thanks to the editor and the referees for their exceptionally helpful comments during the review process."	"Thank you for your endless requests and revisions that sucked the life out of a poor Ph.D. student."
"Thanks to the person in our research group with whom we had long conversations regarding the subject of this paper."	"Thank you for helping us come up with the critical idea that helped solve our problem but you don't get authorship credit."
"Thanks to the Grant Funding Agency for supporting this work. This work was supported under grant N00014-98-7994."	"None of the money was actually used for this paper but we needed to say this in order to get more $$ from them."
"Thanks to certain Professors in other Institutions for their insight and helpful feedback provided at Conference discussions."	"Your comments were not helpful, but I'm writing this in case any of you happen to be reviewing this paper."
"Thanks to our collaborators who provided the data used in this work."	"Actually, we really do want to thank them. They saved us a TON of work! (But they still don't get authorship credit)."

WWW.PHDCOMICS.COM

5.11.1. Author Contribution Statement

Some journals include an additional section before or after the acknowledgments called "Author Contribution Statement". Note that this is different from the footnote that is usually used to document shared first authorship of a manuscript (see Section 5.2.3). In the author contribution statement the authors are supposed to describe in more

Chapter 5. The Structure of a Scientific Paper

detail what has been their contribution to the work: "*A did the experiment, B performed the analysis, etc..*"

The author contribution statement is quite a new thing and I suspect it has been included partially to help young scientists to document their contribution and get more credit for their work (which is a good thing) and partially because there have been some serious cases of very prominent scientific fraud in recent years. This is discussed briefly in Section 5.13. The author contribution statement makes it easier to trace the responsibility for the work presented by individual authors.

5.12. Citations/References

Citations also referred to as references are a very important part of your paper. They should provide the reader with an insight into the state-of-the-art of the field you are working in. A discussion on how to find relevant citations is included in Section 5.5.2. Here are a couple of further important points:

- Cite yourself! This is important first and foremost because you probably already contributed to the state-of-the-art in the field. Further it is important in order to improve your citation index (see Sections 2.6 and 2.7). On the other hand - don´t overdo the self-citations. I have come across papers in which more than 80% of all citations were self-citations. They did get away with it, but it does not convince the reader that this really is a field of general interest. As a rule of thumb, you should not have more than 30% self-citations.
  - It is not a problem to cite very many of your own papers if there is no limit on how many papers you can cite in total. Just make sure you do not exceed the 30%.
- Cite several papers from the journal you want to publish in. This is always a good argument when you write the cover letter to the editor as to why your paper is particularly relevant for his/her journal (see Chapter 12).
  - If possible, cite papers from the journal you want to publish in from the two previous years. That way you contribute to the journal's impact factor (see Section 2.4).
    * If for example the journal you want to publish in has an impact factor 2 and you cite two papers from the two previous years, then your paper is at the very least "impact factor neutral" for that journal. The editor will know that the journal can never "lose" by publishing your paper.

- Make sure that you cite several recent papers, to emphasize that this paper is on a "hot" topic.

  – If you want to publish in a really high-ranking journal like *Nature* or *Science* or similar, you should make sure to cite several recent papers in similar high-ranking journals to emphasize that this is a "hot, hot" topic.

- In case you only have rather old citations, it can be a good idea to explain the reason for this in the introduction. For example that a method has recently become available making it possible to address this old problem again in a new way.

  – The idea is to convince the journal that you are creating a new "hot" topic.

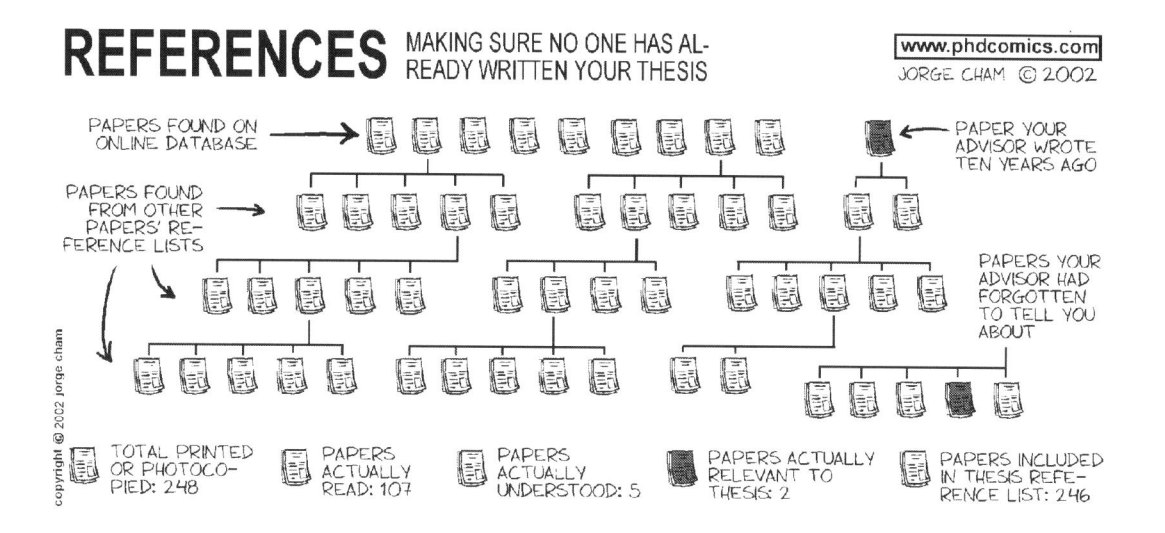

![REFERENCES - MAKING SURE NO ONE HAS ALREADY WRITTEN YOUR THESIS. www.phdcomics.com JORGE CHAM © 2002. Papers found on online database; Papers found from other papers' reference lists; Paper your advisor wrote ten years ago; Papers your advisor had forgotten to tell you about. Total printed or photocopied: 248; Papers actually read: 107; Papers actually understood: 5; Papers actually relevant to thesis: 2; Papers included in thesis reference list: 246.]

A final comment: You will often use some of the same citations in different papers. It is therefore a good, time-saving idea to keep your own citation database. I usually write in *LaTex* and I use the program *JabRef*, which is free. If you want to write in *Word* you have to use something else, for example the commercial *EndNote*. It is possible to export references between the two.

103

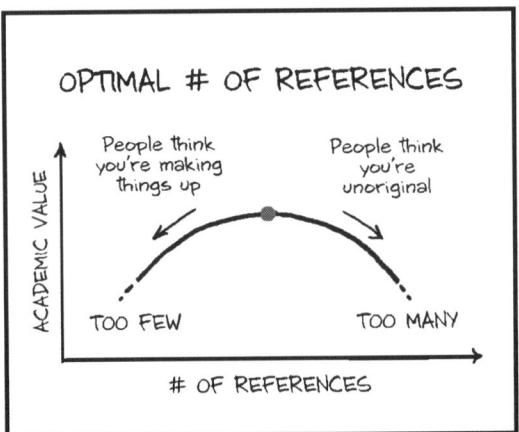

Chapter 5. The Structure of a Scientific Paper

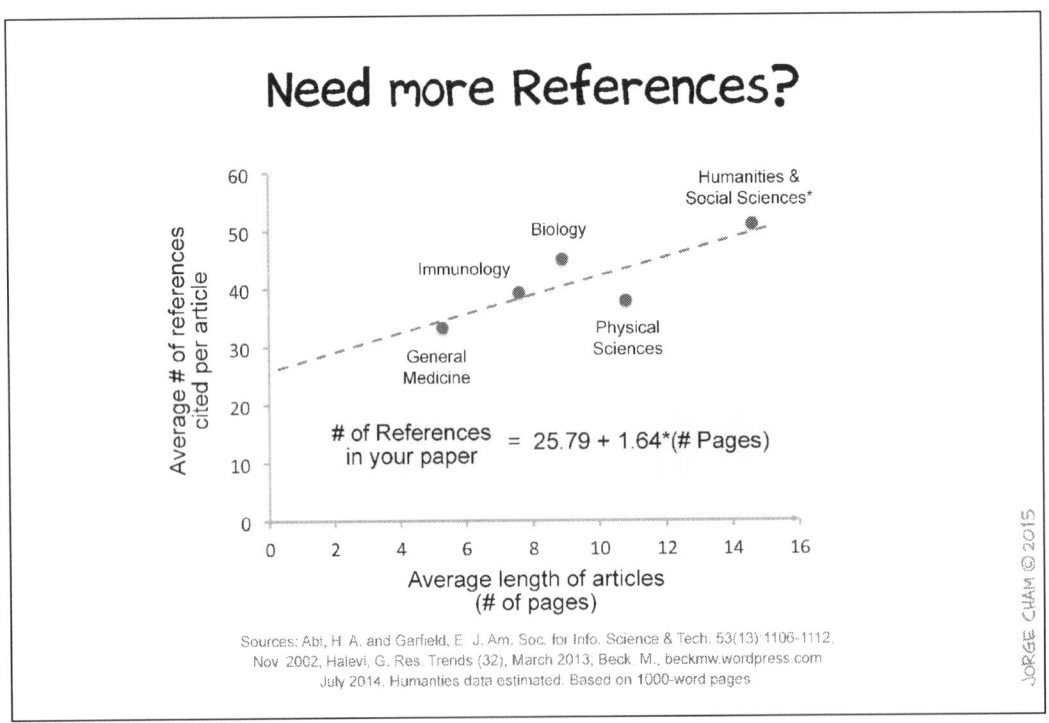

5.13. Appendices/Online Supplementary Material

An appendix is an additional section in a paper included at the very end after the acknowledgments or the citations. An appendix is used to provide information that is important for the paper, but which is so detailed that it would distort the information flow if it was included in the main paper. An appendix can contain for example a detailed description of calculations, additional data sets, more detailed explanations of certain aspects of the methodology, etc..

In former times, when the scientific community was much smaller and people wrote fewer and longer papers, appendices were quite popular. Later, the trend moved towards short papers and the appendices disappeared almost completely, which is too bad, because they can contain very valuable information. Most journals still accept appendices as long as it does not lead to the paper exceeding a maximum allowed length.

In recent years the appendix is experiencing a renaissance in the form of "online supplementary material", which is simply an appendix that is not printed, but only stored electronically. Even journals that only exist online, such as *Scientific Reports* from *Nature Communication Group* use "online supplementary material".

At the moment online supplementary material is provided on a volunteer basis. You decide yourself if you want to provide additional information or not. When at all relevant, do provide online supplementary material because it shows that you are thorough, which makes a good impression on the referees and also makes your paper easier to read.

There are moves in the scientific community towards an extended, obligatory sharing of data, the idea being that scientific data generated should be available to the whole scientific community. It has even been suggested that this idea should be extended to computer codes used to generate data in computer simulations so that others can use the code and also check it for errors. Many scientists already make their computer code publicly available or supply it upon request, but for various (obvious) reasons there is some reluctance in the community to make this a general rule.

5.13.1. Scientific Fraud - Stopping Schön & Co.

There is an additional, not so nice but reasonable incentive for providing more scientific information online: The detection of scientific fraud. Scientific fraud is nothing new. There are many examples of this, one of them the famous Piltdown man fossil "discovered" in 1912 and exposed as a fake in 1953. However, the problem seems to be increasing, which is hardly surprising seeing how science is becoming more and more competitive, involving more and more people, and with more and more money at stake. One of the most prominent cases in recent years is ex-professor Dr. Hwang Woo Suk from Seoul National University. In two articles published in *Science* in 2004

5.13. Appendices/Online Supplementary Material

and 2005, he reported the creation of human embryonic stem cells by cloning. Not only did it turn out that Hwang had committed serious ethical misconduct by using eggs from his graduate students(!!), his human cloning experiments were also found to be fraudulent and the papers have been withdrawn.

Another prominent case is from the late nineties of the last century. The ex-Dr. Jan Hendrik Schön was working at the time at Bell Labs, one of the most famous research institutions in the USA (the transistor was invented there). Schön worked in molecular electronics, which is not so far from my own field of surface science; I clearly remember following his amazing success with numerous *Nature* and *Science* papers, asking myself, as I was struggling with the realities of experimental work and the stress of not having a permanent or tenured position: *"How does he do it? How on earth does he do it?"*. Well, the truth was that - he did not. He did not even do any measurements; his results were all made up! After a while people did start to get suspicious and when Lydia Sohn at Princeton University discovered that two data sets published in *Nature* and *Science*, supposedly measuring completely different things, had identical noise distributions, the scandal started to roll. A committee later found out that there were serious issues already with the data in Schön's PhD thesis and the University of Konstanz revoked his PhD in 2004.

I must confess that when the fraud was discovered I was extremely happy. I ought to have been sad that all those exciting results were not true, but I wasn't. Amazingly Schön was never taken to court, and the case did not have any consequences for any of his co-authors either, including the director of Bell Labs, who had been very happy to travel around the world presenting Schön's exciting data.

The real tragedy of the case I learned a few months later, when I visited a conference in Vienna and spoke to a very nice PhD student from a very famous American university. I shall never forget the look in his eyes when I asked him: *"Have you heard about the Schön case, isn't that funny - ha ha..."* and he replied *"I hate that man, he cost me two years of my life. For two years I tried to reproduce his results. I went to my supervisor and said to him, it is impossible, it cannot be done, but all he said was: These results are published in Nature and Science, get back to work"*.

For a very interesting read on the whole Schön case see [47].

Chapter 6.

Drafting the Manuscript: A Possible Approach

The problem with paper writing is that unless you need to meet a deadline for a conference, you can always postpone it. **Paper writing is always important but never urgent and this poses a great danger.** Therefore the first rule for drafting the manuscript is to allocate time for doing so: fixed time slots every week, which you do not use for anything else. It can be a good idea to work at home if possible to avoid disturbance. For an extensive discussion of this, see Chapter 4.

In the next section I describe a procedure for structuring the paper writing process. This is the procedure I use myself and many students have told me that they find it very helpful. An alternative approach is presented by George Whitesides in [11]. George Whitesides is one of the world's most prolific chemists. The two approaches are quite similar, but Whitesides uses an outline approach as a tool for communicating between supervisor and student.

The key point in the procedure outlined below is that you do not do any writing before you have done some data analysis, decided what "the take-home message" of your paper is going to be and prepared the first drafts of figures and tables demonstrating the validity of the "take-home message". In other words - you do not start actually writing before you are reasonably sure about what you are going to write. This may seem obvious, but I have found that surprisingly, people often get bogged down while writing simply because they do not know what they actually want to write.

6.1. Writing the Paper Step by Step

- The very first step is to get the ideas for your experiments/theoretical work. This includes realizing that some unexpected results you have got should be pursued further (tip for the experimentalists – try to do preliminary data analysis along with the experiments because it is very annoying if you realize later that crucial results are missing). Discussion with colleagues is vital! Don´t be afraid of bouncing your ideas off other people – just make sure they don´t steal them!

  - The point about extensive discussion of results and ideas is really paramount. Scientific work is a creative process and part of that creative process is discussing with other people. Thus, an environment in which you can discuss freely is very important. A key factor for a good discussion environment is that nobody is afraid of seeming stupid.

    * If you are a supervisor, do make sure that your students have the feeling that they can come to you with all their questions and issues. Take your time to talk to them. Very often students will come to you with a problem and during the process of explaining the problem to you, they come up with the solutions themselves; all you have to do is listen. Interestingly enough, you must truly listen; it is not enough to pretend that you are listening! That will not produce the desired effect. I am speaking from experience! Also, if the students ask you questions

Chapter 6. Drafting the Manuscript: A Possible Approach

that you do not know the answer to, do not try to be evasive. It is much better to admit it up front if you do not know something. Clever students will spot any bluff immediately and lose respect, not-so-clever students will get confused.

* If you are a student and do not have a supervisor whom you can talk to about your work as much and as freely as you would like and if you are working largely on your own so that there is nobody else readily at hand: Do not wonder if you are feeling that you are not making any progress! That is perfectly normal. Discussion is necessary. Create a discussion environment yourself with some of your fellow students.

  · An anecdote from my own time as a student illustrates the above. For the first three years, I studied mathematics and physics in parallel as was usual in Copenhagen at the time. In my second year, we had a professor who came across as very introverted and dry (quite typical for a mathematician, you might say), although he was a brilliant teacher. At some point he was presenting a proof to us (I have completely forgotten what about.). As a side remark, he mentioned that this proof was really an astonishing achievement, because the person who had come up with it had served a life sentence in prison and so he had never had the possibility of communicating with other mathematicians. The professor made the comment that he thought it truly remarkable that you could come up with something like that completely on your own. With that, the story would have ended had not the audience insisted on being told why the person was in prison in the first place. It turned out that the poor guy had suffered shell shock in the WWI trenches and upon returning from the war, had killed his wife and children. Note! My mathematics professor (who I had thought never spoke to anybody) did not think it strange that a person who has killed his wife and children can do mathematics; he thought it strange that you can do mathematics without discussing with others.

* Remember that your brain works best if you are not stressed. There are may famous anecdotes about famous scientists getting their breakthrough ideas in their sleep, in the bathtub, lying under an apple tree, etc.. The moral of these stories is that it is important now and then to rest under an apple tree after a nice warm bath.

6.1. Writing the Paper Step by Step

Where my best ideas usually come from

[Bar chart by Jorge Cham © 2013, www.phdcomics.com, showing "IDEAS" on the y-axis with bars for: SITTING IN MY OFFICE, DOING THE DISHES, WALKING TO LUNCH, TALKING TO SOMEONE, TAKING A SHOWER, DOING TWO OR MORE OF THESE THINGS AT THE SAME TIME.]

- While doing the data analysis, you should start thinking about what is the main conclusion, the "take-home message" of your paper (see Section 4.2). Remember: A paper should tell a concise story. You may need further experiments/calculations to support the "take-home message" so it is important to think about this at an early stage without losing an eye for unexpected effects. If we knew the results beforehand, it would not be science. Also, perhaps it turns out that you have more than one take-home message and it would be better to write two papers instead of one (see Section 4.3).

- Decide what figures and tables you want in your paper and prepare first draft versions. Figures and tables present your results, they are your way of convincing the reader that your "take-home message" is valid.

    - **Once you have decided on the figures and tables, the paper is actually about two-thirds finished even though you have not written a single line of text yet.**

Chapter 6. Drafting the Manuscript: A Possible Approach

- Decide what journal you want to submit your paper to. Check the profile of the journal, which is usually available online. It will often make sense to submit to a journal that contains many papers of relevance to your work. You will naturally cite many papers from that journal in your new paper, hence improving the journal's impact factor if they accept it (see Sections 2.4 and 5.12). Look at the papers you have needed for your data analysis/background research.

- Write a first version of the abstract and make up a preliminary title – this sharpens your mind for what you really want to say with the paper. See Chapter 5.3 for a detailed description on how to write the abstract.

    - Some literature on paper writing suggests that you write the abstract at the very end. Of course this is entirely up to you, but I find that it is actually very useful to write the abstract first because it forces you to clarify what the paper should really be about, what is the "take-home message".. As mentioned earlier, I have found that people surprisingly take a very long time writing a paper, simply because they do not know which paper they are writing. **When you have written the abstract, you know which paper you are writing.**

- Write a first version of the methods, results and conclusion sections. I usually write this in parallel starting by writing down loose sentences with important points for each section and then putting them together in the end. If you are not a native English speaker, don´t worry about the language. Write the ideas as they come to you, in your native language, in incorrect English; mix the words from the two languages if needed, just get something down onto paper.

    - Note: Mixing words is OK, but I do not recommend writing the whole paper in your native language and then translating it. This is not likely to yield good results.

- Then write a first version of the introduction. Some people also write the introduction parallel to the sections above; this is a matter of taste. If you have already written several related papers, then it might be easier to write the introduction at an earlier stage.

- Give your paper to co-authors for comments; **all co-authors must approve the final version.**

6.1. Writing the Paper Step by Step

– When you are giving the paper to co-authors for comments, set deadlines for when they have to return their reply: "*If I do not hear from you within the next three weeks, then I will assume that you are satisfied with the paper. Please report back to me if you are not able to answer within this period of time*". This can speed up the process immensely.

- Make sure that somebody with very good English proofreads the paper. Your English does not have to be perfect when you submit to a scientific journal, but it must be clear to avoid any misunderstandings. It should also be devoid of most grammatical errors. Sloppy English may well create the impression of "sloppy science" with the referees.

    – I do not generally recommend professional translation offices. They are very expensive and unless they are really specialized in your particular topic-area, they may actually make mistakes. If you feel you need extra assistance, the best is to find a person you can interact closely with so that the correct scientific idea is conveyed. Tip: Try to find a native English-speaking undergraduate student in your field to proofread the paper for a modest fee and perhaps a mentioning in the acknowledgments. In Europe you can try to find an English Erasmus Student.

    – Spell checking is not enough. When my supervisor returned the first draft of my PhD thesis, he said that unfortunately (because he really appreciated

Chapter 6. Drafting the Manuscript: A Possible Approach

them) but given that this was not a theology PhD and in the interest of correct science, I would have to replace my *"incident angels"* with *"incident angles"*.

- Make sure that the person proofreading your paper does not have the same native language as you do (unless you are both English speakers). We all make language-specific mistakes, which other people with the same native language will not pick up easily, even if their English is considerably better than ours.

- In case you wondered - the reason papers in journals like *Nature* and *Science* all read so well is because they are rewritten by the editorial staff. I was reminded of this when we published a technical comment on ancient plant fibers in *Science* [43] (see also Section 5.2.1). The authors of the original paper were very late in submitting their reply to our comment; so in the end it was seven months after submission before I received the proofs and swelled with pride over my excellent English. The swelling lasted for about 10 minutes. Then reality struck. I received a second email with the original manuscript in much less excellent English and a note asking me to check for any unwanted changes in the scientific argumentation caused by the editing process.

• Repeat the last steps a number of times until the paper is deemed finished by everybody. Remember, the perfect paper has never been written.

6.1. Writing the Paper Step by Step

6.1.1. A Small Reflection on the Recycling of Text and Figures

Let us get this straight: In principle all the work that you publish in a paper, be it text or figures, must be new and original. In practice however, there is no need to reinvent the wheel more often than necessary. As long as you present original data in your paper you are not likely to get into any trouble provided you take a few precautions. If you have made an excellent figure of an experimental setup that you have used for several different experiments, this figure can be reused essentially as it is in the various papers, without any referencing. It is enough if you make minor modifications, such as changing the size of an arrow or adding or removing minor components. Then it qualifies as a new figure.

When it comes to text, the same thing applies. If you have written an excellent methods section in one paper and you are using the same method in another paper, you can reuse the old methods section if you make some minor changes to the text. It would not be a good idea just to refer to the other paper, because it is inconvenient for the reader not to be able to follow the experimental procedure directly.

Whatever you do, you should only reuse your own text and figures. **Never, ever reuse other people's text and figures without proper referencing and official permission.** It will probably not be a problem for you now, but one day, when you are rich and famous and/or Minister of Defense in Germany (see Section 6.1.2), it might come back to haunt you!

- **BE AWARE:** It is your responsibility as an author and NOT the responsibility of the journal to make sure that your paper is "legal" in every respect. This will be clear from the copyright form you need to sign before your paper is published.

- Getting official permission for recycling a figure is actually often very easy. Only very rarely will you need to write the authors directly and ask for permission. Very many journals and text book publishers have a general permission for free use of all illustrations as long as this is for academic purposes and properly referenced. This will be stated on their web page.

    – Sometimes you may want to reuse a figure in a modified form. In this case the proper way to reference would be to write in the figure caption: "*modified from X*".

- If you need to reproduce photographs, you should sometimes be particularly careful even if it is your own photograph.

Chapter 6. Drafting the Manuscript: A Possible Approach

- If you want to include a picture of a unique object that does not belong to you or your group, (i.e. a painting or an archaeological sample or something similar), you must obtain explicit permission from the institution that owns this object. They may also insist that you use their official photograph of the object and they may demand money as well, even if it is for academic purposes.

- If you want to include a photo taken by somebody else, this person must be referenced by name in the figure caption even if it is published in a journal with general permission to use illustrations for academic purposes. The book/journal where you got the illustration from must of course also be referenced.

6.1.2. The Guttenberg Case - "Dr." von Copy zu Paste

The Guttenberg case is a remarkable, recent case of academic fraud, which led to the abrupt end of the political career of one of the most prominent young, German politicians: ex-Dr. Karl-Theodor Maria Nikolaus Johann Jacob Philipp Franz Joseph Sylvester Freiherr von und zu Guttenberg. At the time when the scandal hit, Guttenberg was the German Minister of Defense. Handsome, young, heir to a family fortune, of ancient German aristocracy and equipped with a very classy wife (a descendant of Bismark on her father's side, but with the looks of her Swedish mother). At the time, Guttenberg was considered one of the most likely people to be elected as the next Chancellor candidate for the German Conservative Party following Dr. Angela Merkel (yes, she is genuine). Then, in 2011 a professor of law at the University of Bremen, Dr. Andreas Fischer-Lescano, discovered that a newspaper article was included in Guttenbergs PhD thesis without proper referencing. More unreferenced citations were found and eventually the University of Bayreuth appointed an official investigation commission. The process escalated through the website "Guttenplag" (http://de.guttenplag.wikia.com/wiki/GuttenPlag_Wiki), on which everybody was invited to upload information about unreferenced citations in the Guttenberg thesis. The current status is that there are unreferenced citations on 371 of 393 pages. The University of Bayreuth revoked Guttenberg's PhD and shortly thereafter, he resigned. Since the Guttenberg case, more politicians have been forced to retire because of fraudulent thesis work, including the then German minister of Education and Research! ex-Dr. Annette Schavan. I find that it actually speaks a lot for Germany that a PhD is considered important for a political career.

Chapter 7.
Language: Dos and Don'ts

Clear, precise language is important for any scientific paper. Your language does not have to be perfect, but there are certain rules that should be adhered to, rules that are often broken, both by native and non-native speakers. In this chapter I outline some of the most important aspects to be considered. One issue you should consider before anything else: Even though a very important part of a paper is to place your results in a scientific context, which means comparing it with the work of other people, utmost care ought to be taken not to insult or criticize other people's work. This is discussed in more detail in Section 5.5.2.

Chapter 7. Language: Dos and Don'ts

> **Sentences you will probably never read in a published paper:**
>
> "We were totally surprised it worked!"
>
> "We just thought it'd be a neat thing to do."
>
> "I'm only doing this to get tenure."
>
> "Oops."
>
> "Previous work by XXX et al. is actually pretty good!"
>
> "To be honest, we came up with the hypothesis *after* doing the experiment."
>
> "The results are just 'OK'."
>
> "Future work will... ah, who are we kidding? We won't get more funding to do this."
>
> JORGE CHAM © 2010   WWW.PHDCOMICS.COM

7.1. The Danger of Synonyms

At school we are taught to write essays in our native language. In all the languages I have come across so far, this involves using synonyms to vary the language. Synonyms are words that mean the same in a particular context: article and paper, device and instrument, speed and rate, method and model, etc..

7.1. The Danger of Synonyms

We are taught at school that a repetition of the exact, same words in several sentences following each other is not elegant. Be careful, however: Synonyms might be good in fiction, but they are confusing in science. Do not use synonyms when you write a scientific paper. Your prime aim is clarity and NOT elegance. An example:

- *"Sugar levels of up to 20% were observed. The experiments were carried out at the maximum sugar concentration."*

    – Level and concentration are here used as synonyms. For somebody not in the field or not so strong in English, this is confusing.

- What you should have written is: *"Sugar levels of up to 20% were observed. The experiments were carried out at the maximum sugar level of 20%."*

    – Yes it is a bit clumsy, and perhaps you could reformulate it, but this should not be done by using synonyms. As it stands it may not be elegant, but it is clear.

- Not only avoid synonyms, make sure that you always refer to a particular concept in exactly the same way. If you take a look at the abstract *"The Quality of Sweet Pastry can be Determined by Fruit Fly Spotting"* in Section 5.3.3, you will see that the expression *"sweet pastry"* is used rigorously throughout the abstract. This would have continued throughout the rest of the paper, were it to have been written. Under no circumstances should you change horses in mid-stream and suddenly refer only to *"pastry"*, for example.

- The use of synonyms and the urge to vary the language when we write is so inherent that it is almost impossible to avoid. The last stage when I write a paper is always going through it and looking vary carefully for any synonyms or changes in crucial expressions. Even so, more than once has it happened that I have only spotted them after the paper has been published.

- Finally you should make sure to use the same label for an issue throughout the paper. A common mistake: using different units for comparisons, e.g.:

    – *"The resolution of the sensor is 0.1 nm. This is negligible compared to a vibration amplitude of 0.2 µm."*

        * The sentence is true, but if you just read it quickly you may get irritated because "0.1" is not negligible compared to "0.2".

Chapter 7. Language: Dos and Don'ts

7.2. Bigger, Better, Many

Science is about precision. Avoid "relativating" words on their own except perhaps in a catchy title. These words compare the object they describe with something else (large, big, several, typically, etc.). The following examples from drafts from some of my own publications illustrate this:

- *"A crystal cut with ultra high precision was used for the experiments."*
  - What does *"ultra high precision"* mean? How accurately was it cut? At the very least better than $0.25°$ most likely, because this is a typical industrial standard now, but what is somebody reads the paper in 20 years when the industrial standards might well have changed? What would they think *"ultra high precision"* means?

- *"The miscut of the crystal was considerably less than $0.25°$."*
  - Better than the first sentence, but still not ideal. What does *"considerably"* mean? Either you can quantify it or you cannot.

- *"A large number of data points were collected."*
  - How many? The meaning of *"large number"* varies drastically from topic to topic.

- *"The crystal was heated to about 500 K."*
  - Error bars?

- *"The crystal was heated to 500 K."*
  - Even worse. Strictly speaking, an expression like that must be interpreted as follows: The measurement is accurate to within $+/- 0.5°$, but often this is not what is meant. Always supply error bars.

There can be situations in which less precise statements are OK. For example in the introduction "softening" or "enhancing" statements can be useful:

- *"The diameter of the skimmer is typically in the range 1-400 µm."*
  - The word *"typically"* softens the statement. Even a referee who has used a 500 µm skimmer all his life cannot object.

- "*Quartz is widely used in industry.*"
  - The word "*widely*" enhances the sentence. It serves to justify the importance of the work. Ideally it requires supportive facts i.e. examples of applications with references or similar. For example a statement such as: „*The annual production amounts to . . .*" supports the claim "widely used". Authors often omit such non-scientific references, particularly in highly specialized journals. But it makes a good impression if you include them.

7.2.1. Avoid Words with a Specific Meaning Unless you Use Them to Mean Just That

Do not use words from statistics unless you are describing statistics. For example:

- Do NOT write: "*There is a significant increase*" unless you perform the statistical analysis showing that you really do have an increase which is significant in the statistical sense.

Instead write:

- "*There is a clear increase.*"
  - You should still provide an explanation. "*clear increase*" is not clear enough, but it can be more "hand waving" because "*clear*" does not have a specific, statistical meaning.

7.3. Active or Passive Voice?

- "*In this paper we present experiment X*" (active voice)
- "*In this paper experiment X is presented*" (passive voice)

The use of the active or passive voice is one of the issues that tends to occupy a writer's mind more than is actually necessary. In fact, there are no rules here. It is simply a matter of personal taste; you can choose between the two as you prefer and you can even change from one to the other when you move between sections in the paper. Some people have the (wrong) idea that writing in the passive form is more sophisticated and shows greater learning. However, for a scientific paper, style is not inherently important. What matters is that your scientific "take-home message" is communicated to the reader as clearly as possible. The truth of the matter is that

Chapter 7. Language: Dos and Don'ts

the active form is easier to write AND also easier to read. The latter is not unimportant. It's essential to remember that **we do not write our papers primarily for native English speakers**. If you check into this, you will see that the active form is largely preferred by all the major scientific journals such as *Nature*, *Science*, etc..

- The passive form can be used in the methods and results sections:
    - *"The pH-value of sample X was measured"* (passive voice)
    - *"We measured the pH-value of sample X"* (active voice, but not so good – it sounds as if all authors were hanging around with a piece of litmus paper!)

- The active form should be used for the expression of personal opinions/beliefs:
    - *"We have not found any references in the literature treating the general case."*
        * Alternatively, if you really hate using personal pronouns, you could write:
    - *"The authors have not found any references in the literature treating the general case."*
        * Both these sentences would "stand in court". In the worst case, the referee could point out a reference, but that does not make the sentences wrong, because it remains true that the authors did not manage to find any references.
    - *"No references have been found in the literature treating the general case."*
        * This sounds as if the whole of humankind has been hunting for references.
    - *"There are no references in the literature treating the general case."*
        * This sentence can be proven wrong by the referee or others providing a reference and should therefore be avoided.

7.4. Present or Past Tense?

This is an issue which often leads to confusion, but actually the rules are quite simple and can be listed as follows:

- **General scientific conclusions MUST be kept in the present tense, because they are always true:**
    - "*The measurements showed that the a-quartz (1000) crystal surface has a hexagonal structure.*"
    - "*The measurements have shown that the a-quartz (1000) crystal surface has a hexagonal structure.*"
    - Of course you can also write: "*The measurements show that the a-quartz (1000) crystal surface has a hexagonal structure.*"
        * BUT NOT: "*The measurement showed that the a-quartz (1000) crystal surface had a hexagonal structure.*"
        * This would imply that the properties of α-quartz have changed since the experiment, which cannot be the case.
        * There might be a bit of confusion here for non-native speakers of English. Some other grammars may be more critical about mixing the tenses, but in English, this is definitely possible.

Chapter 7. Language: Dos and Don'ts

- The descriptions of the methods should be kept in the past.
  - *"The crystal structure was measured using He-scattering."*
    * NOT: *"The crystal structure is measured using He-scattering."*. This sounds as if we are still measuring, which is clearly not the case.
  - Possible exceptions by which methods can be described in the present:
    * When a theoretical methodology is explained, then it can be kept in the present. This is the Socratic method – take the reader through the proof.
    * When you describe a „recipe": *"1. Take three eggs . 2. Mix them with two liters of milk ..."*. If you recipe is presented as a result, then it MUST be written in the present.

I suggest that you go back and look at the three abstracts in Section 5.3.3. Note how the results are always presented in the present tense and how that makes the result sentences "stand out" in the abstract. It is important to get this right.

DECIPHERING ACADEMESE

YES, ACADEMIC LANGUAGE CAN BE OBTUSE, ABSTRUSE AND DOWNRIGHT DAEDAL. FOR YOUR CONVENIENCE, WE PRESENT A SHORT THESAURUS OF COMMON ACADEMIC PHRASES

"To the best of the author's knowledge..." = "WE WERE TOO LAZY TO DO A REAL LITERATURE SEARCH."

"Results were found through direct experimentation." = "WE PLAYED AROUND WITH IT UNTIL IT WORKED."

"The data agreed quite well with the predicted model." = "IF YOU TURN THE PAGE UPSIDE DOWN AND SQUINT, IT DOESN'T LOOK TOO DIFFERENT."

"It should be noted that..." = "OK, SO MY EXPERIMENTS WEREN'T PERFECT. ARE YOU HAPPY NOW??"

"These results suggest that..." = "IF WE TAKE A HUGE LEAP IN REASONING, WE CAN GET MORE MILEAGE OUT OF OUR DATA..."

"Future work will focus on..." = "YES, WE KNOW THERE IS A BIG FLAW, BUT WE PROMISE WE'LL GET TO IT SOMEDAY."

"...remains an open question." = "WE HAVE NO CLUE EITHER."

JORGE CHAM © 2004
www.phdcomics.com

7.5. Keep It Short and Simple (Kiss) - A Note Particularly for German Native Speakers

Modern written scientific English is very similar to spoken scientific English with short sentences with two subclauses at the most and never more than two lines long. In written scientific German, there is a danger that the writer might be considered unintelligent if the sentences are too short. This is not so in English. If you write very long sentences, people are not going to think that you are particularly clever, they are just going to think that you are German.

Chapter 8.
A Paper or a Patent or Both?

From a scientific point of view, a patent can never count as much as a paper, simply because the authorities approving patents are not judging the scientific importance of your work. Their only concern is if the idea presented can be considered novel enough to merit a patent. Furthermore, although you are supposed to write how you technically realize your idea in a patent, it is perfectly possible to submit patents with ideas that have not been realized in practice. Big companies sometimes do this deliberately to confuse the competition.

So to stand best in the scientific community (see Section 2.3), you should always aim at presenting your work in scientific papers, but this does not mean that you cannot or should not submit patents as well. You just have to do things in the right order:

- If you want to patent an idea you have to submit the patent first, before you submit a paper.

- My way is to first write the paper using the standard paper structure (see Chapter 5) and then give that paper to a patent lawyer who will translate and restructure it into patent language.

- When the patent process is finished (that is when an official submission confirmation from the patent office is available), I submit the original paper.

    - The reason it has to be done this way is that patenting follows the rule "first to file", which means that the first person to submit a patent on a certain idea will be granted the patent unless the idea has previously been published elsewhere, i.e. by you in a scientific journal.

    - Be careful: Conference presentations also count as published, even if you do not submit a paper to the conference proceedings but just present a talk or a poster.

Many universities have offices advising on patents. In most places you are legally obliged to hand over descriptions of every invention which you carry out as part of your work and which you wish to patent. The university authorities will then decide if they want to patent. If yes, it means that the university will cover all patenting costs. You (the inventor) will typically get 1/3 of the income generated from the patents. This is at least the rule for most places in Europe. The university authorities may decide not to patent your idea, in which case you have the right to do so yourself. This means covering all costs yourself but also retaining all the rights.

Chapter 9.

The Random Paper Generator - No this is NOT an Easy Way Out

Scientific papers are written according to an established, well-defined structure as described in Chapter 5. In 2005 Jeremy Stribling and student colleagues at MIT decided to exploit this for a bit of fun and wrote *SCIGen: The Random Paper Generator* (http://pdos.csail.mit.edu/scigen/).

In 2007 a group of students under the pseudonym R. Mosallahnezhad succeeded in getting a paper generated using *SCIGen "Cooperative, compact algorithms for randomized algorithms"* accepted for publication in the *Elsevier* journal *Applied Mathematics and Computation* [48]. By now the paper has been removed from the journal website. The journal still exists. In 2014 it had an impact factor of 1.551. I do not know if any changes have been made to the editorial board.

The paper by R. Mosallahnezhad was only removed when the conscientious students who had submitted it made the journal aware of the hoax. In 2012 the French computer scientist Cyril Labbé started to catalog *SCIGen* papers. He uses a program he has developed to automatically detect manuscripts composed by *SCIgen*. So far his program has detected more than 120 conference proceeding papers published by *Springer* and *IEEE* [49]. Thus *SCIGen* papers cannot be recommended as an easy way to boost your paper portfolio.

Part III.
Submitting the Paper

Chapter 10.
Preparatory Steps

10.1. Consider the Psychology of the Referee (and the Editor)

At the end of the day, getting a paper published boils down to convincing a few people, i.e. editor(s) and referee(s) that your work is good. Ironically these few people are the only people who will not read your paper out of pure academic interest, but because it is their job to do so. Most scientists will agree that being a referee is part of our job (see Section 1.2), but it is time consuming and the requests always come when you are busy with other things. (You are always busy with other things.) So unless you happen to referee a paper which is exactly related to your own work (a paper you would have read anyway), there is always a certain energy barrier to overcome. This is important to bear in mind. **You cannot expect referees to read and review your paper with the same passion and love that you wrote it with. You also cannot expect referees to be all-knowing, completely impartial and objective, because no human being is.** Particular psychological points to consider are listed below:

- Referees are human beings. Human beings are sometimes lazy or under time pressure.
  - Make life easy for referees. Make sure that you have written a "bang-on" abstract which clearly explains "WHY YOUR WORK IS INTERESTING" and "WHAT YOU HAVE DONE" (should be presented in that order, see Section 5.3)
    * **When the referees have read the abstract they should be convinced that your paper should be published in the journal.**
  - Basically if referees have to spend a lot of time trying to figure out what

Chapter 10. Preparatory Steps

you have actually done, it is not going to make a favorable impression and they are more likely to reject the paper.

- Referees are human beings. Human beings are sometimes vain.
  - Make sure that you cite the work of all the people you can think of as referees in the paper.
    * Nothing makes a referee more irritated than if you have forgotten to cite his/her relevant work!

- Referees are human beings. Human beings do not want to make mistakes.
  - Although the refereeing process is anonymous, it is still not a nice thought for anybody that they might recommend a paper for publishing, which later turns out to contain mistakes, etc..
  - This is where being a Nobel Prize Recipient comes in so handy or, failing that, having Harvard or MIT or Cambridge in your University address. It will not be enough to get your paper accepted of course, but it will help, because the referees will have greater confidence in you to begin with. This is not the way it should be, but it is human. One of the reasons Schön could get away with his fraud for so long (see Section 5.13.1) was his famous institution and famous co-authors.
  - If you do not have the initial psychological advantage of being famous yourself or from a famous university, help by convincing the referees that they are backing the right horse by mentioning cooperation with famous people – name dropping is good and sensible here! Ideally you would like the famous ones to be your co-authors, but if that is not possible, use the "acknowledgments" to thank a famous person for "*useful discussions*" and for "*taking such a keen interest in this work*". Remember to obtain permission from them first - **do not thank anybody without first having obtained their permission for doing so. That is impolite and could backfire.**
    * This is just a little trick you can use if circumstances make it possible. It is NOT a big deal if you do not acknowledge anybody famous.
  - **Make sure that you have done your "scientific homework". If your paper is written in a clear manner with a well researched state-of-the-art section, a good data presentation with "easy to read"**

10.1. Consider the Psychology of the Referee (and the Editor)

figures and tables, a detailed methods section and a balanced discussion - in short, if you have written a good paper, referees will have more faith in your work.

The Piled Higher & Deeper
Paper Review Worksheet

Stuck reviewing papers for your advisor? Just add up the points using this helpful grade sheet to determine your recommendation.
No reading necessary!

Paper title uses witty pun, colon or begins with "On..." (+10 pt)	
Paper has pretty graphics and/or 3D plots (+10 pt)	
Paper has lots of equations (+10 pt) (add +5 if they look like gibberish to you)	
Author is a labmate (+10 pt)	
Author is on your thesis committee (+60 pt)	
Paper is on same topic as your thesis (-30 pt)	
Paper cites your work (+20 pt)	
Paper scooped your results (-1000 pt)	
TOTAL	

Points	Recommendation
< 0	Recommend, but write scathing review that'll take them months to rebuff.
0-120	Recommend, but insist your work be cited more prominently.
>120	Recommended and deserving of an award

JORGE CHAM © 2005 www.phdcomics.com

Chapter 10. Preparatory Steps

10.2. Try to Get to Know the Referees and Editors

In the previous section, it was pointed out several times that referees are human beings. We human beings tend to react positively to things we know. If a referee or editor knows you or one of your co-authors even slightly, he or she is more likely to think favorably about your work. Hence, a very important way to promote your work is by making yourself known in your scientific community (and/or publishing together with famous people!). I do not have direct experimental evidence to support this, but the following does support my view: In a *Nature* paper from 1997 "*Nepotism and sexism in peer-review*" [45] Wennerås and Wold analyzed more than 200 bio-science grant applications to the *Swedish Research Council*. They had to get a court verdict first to get the *Swedish Research Council* to hand them out, but Sweden has very strict laws allowing the public to view official, administrative documents. In the end they won. In most other countries it would not have been possible to get access to this information.

As part of the *Swedish Research Council* reviewing process, each grant applicant was assigned a competence-score by the committee. Wennerås and Wold could demonstrate that there were two factors that strongly influenced this competence score apart from the work actually done: If you were a woman, it decreased your competence score. If you had co-authored a publication with someone on the committee, it increased your competence score.

So how can you get yourself known in your scientific community other than by writing brilliant papers? A very good way is to present your work at conferences. It is preferable to present your work in the form of talks (oral presentations) rather than posters because talks are more prestigious and they attract more attention.

10.2. Try to Get to Know the Referees and Editors

Over the years I have sensed a certain reluctance to give talks, in particular among students, Here are some replies to the standard "talk-objections" summarized below:

- *"You can have a much more fruitful discussion in front of a poster than during the discussion section following a talk."*

    - This is true. However, what will actually happen is that somebody really interested in your work will approach you after the talk and then you will have the discussion that you would otherwise have had in front of the poster. Furthermore, the chances that the people really interested in your work will notice it and get in touch with you is considerably higher if you are giving a talk rather than a poster, especially at big conferences.

- *"My English is not good enough."*

    - Yes, some people struggle with English. On the other hand, even if you do struggle, rest assured that there are people giving talks using English worse than yours! If your talk is well structured and uses good, informative slides in reasonably correct English, then the people that really matter to you, that is the ones in your field, will be able to follow your talk.

    - The main English problem related to talks is usually not grammar but unintelligible pronunciation. If you have this challenge, know that something definitely can be done about it. My advice is that you invest some money in a good speech therapist and not in an English teacher. The English teacher can teach you grammatically correct English and tell you if your pronunciation is correct, but English teachers are not normally specialized in teaching you to pronounce particularly difficult sounds. This is the job of a speech therapist. Speech therapists are mostly known for working with children who have difficulties with speaking, but many also work with adults. Speakers of different languages have problems with different sounds. If you live in a large city, you may even find somebody who specializes in teaching people speaking your particular language.

- *"I do not have any results yet, but my supervisor insists on my submitting an abstract anyway."*

    – This is the one excuse which I can really consider a good excuse even though it is not actually something you need to worry about. I have listened to several talks in which it was clear that the speakers had not managed to get the results they had promised in the abstract. Mind you, many of these talks were still excellent and interesting. Still, if you are at an early stage in your scientific career and giving perhaps your very first presentation at a conference, it is understandable if you feel uncomfortable about giving a talk without proper results. In this particular case it might really be better to apply for a poster.

Having listed the typical main objections that people admit to, here are the two true main objections which people usually hide:

- First main objection: *"I am afraid of being asked a question in the discussion section that I cannot answer."*

    – This appears to be the main reason why people do not want to give talks. It is really a great pity, because it is no problem at all if you cannot answer a question. Firstly, it is important to remember that we are all damaged by school and we tend to associate (consciously or subconsciously) the conference presentation situation with the oral exam situation, (the audience mysteriously then becoming the examining committee...). This is of course completely wrong. A conference presentation is not at all like an oral exam. The audience are not an examining committee who knows everything about the subject. Most likely you will know much more about your particular topic than anybody in the audience. If you do not know the answer to a question, it is not because you are stupid and/or have not done your homework properly, more likely the person in the audience has misunderstood

10.2. Try to Get to Know the Referees and Editors

something and the question is not really relevant for your work. However, if you do not realize this immediately, it still might not prevent the situation from becoming potentially embarrassing. Therefore, it is useful to employ a standard strategy for such situations:

- How to handle a question which you do not know how to answer:
  * What NOT to do:
    - Do NOT look like a scared rabbit.
    - Do NOT fiddle nervously with your laser pointer.
    - Do NOT utter random sentences like "*I do not know.*", "*I am not sure.*", "*I never thought about that.*".
    - If you are a student, do NOT look for your supervisor in the audience with a panic-stricken face, crying silently for help.
  * What TO DO:
    - 1) Stand back calmly and take a deep breath. Remember, you have all the time in the world, there is no hurry.
    - 2) Then adopt a suitably "serious" pose (a hand, elegantly draped on the chin for example).
    - 3) Keep this pose for about 10 seconds (count to 10 slowly in your head). The trick is to create the impression that you are thinking deeply even though in actual fact you are in a state of minor panic with a completely blank mind.
    - 4) Then give the person who asked the question a benevolent smile.
    - 5) Finally, say the following sentence: "*This is a very interesting question. I need to think about it in a bit more detail. I will get back to you in the coffee break.*".
    - 6) Then turn your head to the chairperson with a friendly, relaxed smile and say: "*Can I have the next question please*".
    - 7) It is of course important that you DO approach the person who asked a question in the coffee break. What will typically happen then is that you can resolve the matter very quickly. In the very rare case, where the person has spotted something, you had not thought about BE HAPPY. This is the reason we go to conferences: To get inspiration and discuss our work with colleagues.

Chapter 10. Preparatory Steps

Remember to thank the person, ask if you can mention him/her in the acknowledgments and offer to send him/her the paper once it is published.

* What to do: A Lemma

  · In one of my workshops, one of the participants told that she had been to a conference where a speaker had been asked three questions and he had answered them all with "*This is a very interesting question. I need to think about it in a bit more detail. I will get back to you in the coffee break*". This is of course a bit unfortunate. Luckily there is an easy way to avoid this situation as I learned from a very famous professor from a very famous American university. All you need to do is to plant a question in the audience. Get somebody to ask you a question which you have prepared yourself in advance and of course know perfectly well the answer to. Tell the person to sit in one of the first rows with a good line of sight to the chairperson.

  · This can actually be a nice way of highlighting an interesting aspect of your work, which it was not really possible to enter into during the main talk. Your answer will be so brilliant and so extensive that there will only be time for one more short question, which could then be handled as described above.

- Second main objection: "*I am afraid that somebody will criticize my work.*"

  – Criticism at conferences can be quite hostile, so brace yourself for something tough and do not be afraid. It is amazing with what force and conviction people can express opinions which are completely wrong. Do not fool yourself into believing that just because somebody speaks in a very convinced manner, this means they have actually understood what you are doing. Most criticism can be handled using the method described above, with the phrase "*This is a very interesting comment..*". Always do your best to remain calm and friendly even if the person is being aggressive.

  – Sometime you can experience a peculiar phenomenon, whereby an elderly man in the audience stands up (I deliberately say elderly man, for I have never seen this done by a young man or a woman, for that matter!). The elderly man starts talking at great length in a slightly/completely incoherent fashion. His general topic usually centers on your line of research being

10.2. Try to Get to Know the Referees and Editors

completely useless and how it actually would be much better to pursue this other line of research (typically the field where he has worked for the last 30 years), etc..

* If something like this happens to you, do not take it personally. This is not a personal attack on you, it's something he does all the time at every conference. Sometimes you can even see how the chairperson frantically searches for other people wanting to ask a questions so that he/she does not have to select him. Sometimes you can actually sense a sigh going through the audience, "*Oh NO, not him again...!*". But, if he is the only one wanting to ask a question, the chairperson has to take him.

* In actual fact, it is great to get an elderly man like this because he is likely to use all the question time available.

* What NOT to do:
  - Do not on ANY account enter into any general debate as to why your line of research is sensible. You do NOT need to justify yourself like that. You have been selected to give a talk at the conference. You have been selected by a board of distinguished scientists. There is no need for further justification.
  - 2) Do not under any circumstances, no matter how rude he is, try to interrupt the elderly man. He has years of training and it will not be easy to interrupt him, and any attempt puts you in a defensive position. It it much more impressive if you simply stand there smiling politely and let him ramble. Remember it is the chairperson's job to chair the discussion.

* What to do:
  - When the elderly man has finally finished you smile politely and say in a calm voice, "*Thank you for your comment. I did not quite catch what your actual question was? Could you repeat the question please?*". There will usually be a display of general merriment in the audience at this point.
  - You will now either get a concrete question (most unlikely), which you can handle as discussed above or (much more likely) the elderly man will set off in a second rambling monologue, which you just listen to patiently.

Chapter 10. Preparatory Steps

- When he has finished his second round, you again smile politely and with a condescending smile and a glance at the chairperson you say: *"I am terribly sorry, but I still do not understand your actual question."*.
- At this point the chairperson will usually interrupt and say something along the lines of, *"I am sorry, but we must move on to the next speaker. I am afraid we will have to defer this discussion to the coffee break."*.

An additional way to use conferences to get yourself known in the scientific community is of course to ask good questions yourself.

Chapter 11.
The Submission Process

In this chapter we go through all the formal aspects related to the submission process step by step. Note! **Do not submit the same paper to two different journals in parallel**. This would make you *persona non grata* for these journals if they discover what you have done. You usually have to explicitly state as part of the submission procedure that you have not submitted the paper elsewhere.

- The first step in the submission process is to decide which journal to publish in. This has been discussed in details in Section 3.3.

- The second step is to make sure that your paper is written in a format which fulfills the formal requirements for the journal. All this information will be available online, but the easiest is simply to take a paper already published in the journal and check that your paper looks the same. The particular points to check are:

  - Length of the full paper, some journals have limitations.
  - Length of title and abstract, some journals have limitations.
  - Does the journal allow you to include citations in the abstract?
  - Does the journal request key words or subject classification numbers?
    * If yes, supply some (see Section 5.4).
  - Does the journal require/allow sections with titles?
    * If yes, are there special requirements for the titles of these sections and their order?
    * If no, you can use the section structure suggested at the beginning of Chapter 5.

Chapter 11. The Submission Process

- Have you prepared your figures and tables in the approved journal format?
  * Remember to check that all figures and tables will look OK in the final, published version (see Section 5.8).
- Have you used the approved journal format for referring to figures and tables?
  * Some journals want you to write Figure 1, others Fig. 1. etc..
- Have you used the approved journal format for citations?
  * In some journals citations are listed as [1], [2] etc. Other journals use a format like [Holst 2015].
- Have you listed the citations at the end of the paper in the approved journal format?
  * I very much recommend that you use a citation database like *JabRef* or *EndNote* for your citations. Firstly, you might need the same citations for several papers and secondly, these databases make it possible to change the presentation format of the citations according to the different journal formats. This is potentially very useful if in the end you have to resubmit the paper to another journal.

- The third step is to think about whom you would like to referee your paper.
  - You have the possibility of suggesting people you would like as referees for your paper. Usually it is appropriate to list at least two and at the most four people. Sometimes there is a special section for this on the electronic submission form. If this is not the case, you can always mention it in the cover letter to the editor (see Chapter 12). I always recommend suggesting referees yourself. It can never do any harm and makes a professional impression. A person you suggest as referee should fulfill the following criteria:
    * Be at least a postdoc.
    * Know the field very well as documented in peer-reviewed publications.
    * Have no shared publications with any of the authors on the paper.
    * Be a nice person!
    * It does not harm if the person suggested has published a lot in the journal you want to submit to, but it is not a must. However, if you

want to submit to a high-ranking journal the person you suggest as referee should have some high-ranking publications.

- You do not have to justify your selection of people that you suggest as referees; it is up to the journal to check if they are suitable. However, and in particular for interdisciplinary work, it can be important to have referees with different expertise. It is a good idea then to write that your paper requires expertise in two different fields to receive proper refereeing and list potential referees from both fields stating which expertise they have.

- The fourth step is to think about whom you do not want to referee your paper.
  - It is perfectly normal that somebody may not be suitable to referee your work, either because of animosities or competing interests or both. Some journals have a special section for this in the electronic submission form. Otherwise do mention it in the cover letter (see Chapter 12). You do not have to justify the people you list in any way, you can just write one sentence along the lines of: "*To ensure a good refereeing process, we kindly request that the following persons are not considered as referees ...*".
  - You should not list 20 names, because let's face it: That may actually cover all experts in your field, but up to 3-4 is fine.
    * It may seem a bit strange that you can just exclude people like that but the explanation is simple. Journal editors are well aware of the fact that science is very competitive and that people can be very emotional about their topics. Both are factors which can lead to a biased review and it is not in the interest of the journal to have an excellent paper rejected because they picked the wrong referee. So they are grateful for your assistance.

- The fifth step is to look at the editorial board of the journal and see if there are any members of the board particularly suited for taking a first look at your paper.
  - If the journal has a professional editorial board, you just check if their general scientific background fits.
  - If the journal has an editorial board consisting of active scientists, then you check if any of them has done work related to your paper.
    * The best is of course if you or one of the co-authors know an editor personally or at least has met him or her briefly (see Section 10.2).

Chapter 11. The Submission Process

· Do not forget to cite this editor's work in your paper if at all possible. Remember there is always quite a good scope for broad-range citations in the introduction.

- Some journals operate with a scientific board as well as an editorial board. The scientific board does a first check of the papers that have been approved by the editors. In this case you should check the scientific board, too and write which scientific board member(s) you find most suitable. In some cases you can select scientific board members via the electronic submission form, alternatively you can write the names of your candidates in the cover letter (see Chapter 12).

- The sixth step is to write the cover letter to the editor (see Chapter 12).

- The seventh and final step is to fill out the electronic submission form and submit the paper. This can take some time.

  - Make sure that you have the correct contact details (address, email, telephone) for all the co-authors.

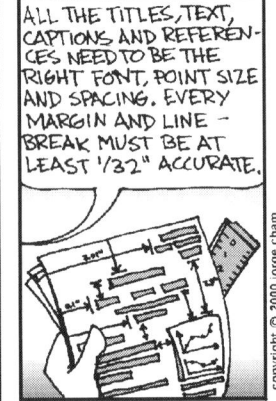

Chapter 12.
The Cover Letter to the Editor

The cover letter is simply a short letter to the editor attached to your manuscript. The reason it is called cover letter is that in the old days (around 20 years ago) when papers were still submitted in paper form, the cover letter would be lying on top of the paper manuscript, covering it. Nowadays the cover letter is usually submitted electronically together with the paper. Sometimes you can upload a cover letter separately, sometime you have to fill out a section on the online submission form called "comments to the editor" or similar. A cover letter should contain the information listed below. I suggest you give the information also in that order. See Section 12.1 for examples of cover letters.

- An addressing of either the editorial board in general or the particular editor you want to look at your paper: "*Dear Dr. ...*" or similar.

- The title of the manuscript.

- A list of all authors as they appear on the manuscript.

- All contact details for the corresponding author.
  - These can be incorporated in the letter head and are not necessary if the journal provides a special "comments to the editor" section on the online submission form.

- A formal request that the paper is considered for publication, stating also, if there is a choice, what type of paper you are submitting, i.e. rapid communication, letter, full paper or review paper. Remember to use the terminology used by the journal to describe the different types of papers.

- A paragraph which is essentially an exciting mini abstract for the paper.

- A brief statement as to why the paper is suitable for publishing in this particular journal.
    - Possible reasons:
        * Your paper fits the general subject area covered by the journal (you may cite from the online description of the journal).
        * Your paper cites several (recent) papers from the journal.
            · This is a good one, it means the journal might improve its impact factor if your paper is published, regardless of how many citations you get (see Section 5.12).
        * Important: If the journal claims to be a high-ranking journal, you have to argue why the paper is suitable for a broad readership.

- Possible additional information:
    - If the journal has a scientific board, you can list here which board members you would like to look at your paper.
    - If you think your results are really nice, suggest to the editor that your paper is taken into consideration as a "cover story".
    - If there is no other place to do so, suggest 3 people you would like to have as referees.
    - If there is no other place to do so, list the people you do not want as referees. You do not need to justify this (see last chapter).

- Finally you can include the names of one or more famous people who recommend your paper to be published in this journal. This is only appropriate for very high-ranking journals and is discussed further in Section 12.2.

- A cover letter should never be more than one page long.

12.1. Examples of Cover Letters

Below you find two examples of real cover letters. Note that the letters are both split up according to the structure given above.

Journal of Microscopy

To the editors of Journal of Microscopy,

Dear Sir or Madam,

Please find enclosed with this letter the manuscript "Imaging with Neutral Atoms - a New Matter Wave Microscope"

Authors: M. Koch, S. Rehbein, T. Reisinger, G. Bracco, G. Schmahl, W. E. Ernst and B. Holst

We kindly request that you consider this manuscript for publication in the Journal of Microscopy as a Rapid Communication.

The paper describes a completely new type of matter wave microscope. The main result is the first image ever obtained using neutral helium atoms.

We find that there can be no doubt that this paper is very suited for the Journal of Microscopy. We take the liberty of suggesting that you consider it for a "cover image".

Yours sincerely

Bodil Holst

Scientific Reports

To The Editorial Board of Scientific Reports,

Cover Letter for Manuscript: Viking and Early Middle Ages Northern Scandinavian Textiles Proven to be made with Hemp. Authors: G. Skoglund, M. Nockert and B. Holst

Dear Sir or Madam,

We kindly request that you consider this manuscript for publication in Scientific Reports.

In this paper we present a textile fiber investigation based on polarization microscopy (the modified Herzog test) of 10 historical textiles from Scandinavia, including the Överhogdal Viking Wall Hanging, the most famous Scandinavian Viking Textile. Up till now they were all believed to be made of flax. It is generally believed that in Viking and early Middle Ages Scandinavia, hemp was used only for coarse textiles (i.e. rope and sailcloth). In this paper we show that 4 of the textiles, including two pieces of the famous

Chapter 12. The Cover Letter to the Editor

> *Överhogdal Viking wall-hanging are in fact made of hemp. In three cases hemp and flax are mixed. Our find suggests that rethinking the organization and resource management of textile production in Viking age Scandinavia is necessary. It highlights the problematic tendency in the archaeological community to identify all plant fibers as flax without proper analysis (as reflected for example in a recent Science article, which we commented on, E. Kvavadze et al., 30,000-Year-Old Wild Flax Fibers, Science 325 1359 (2009), C. Bergfjord et al., B. Holst, Comment on "30.000-Year-Old Wild Flax Fibers, Science 328 1634-b (2010)). This has created a distorted view of textile production.*
>
> *We find that our results are of interest for such a broad readership that they merit publication in a high-ranking journal, such as Scientific Reports. We point out that a previous paper on the identification of textiles published by Holst and co-workers in this journal: "C. Bergfjord et al, Nettle as a distinct Bronze age Textile Plant", Scientific Reports 2, 664, doi: 10.1038/srep00664 (2012) has so far been viewed more than 6000 times.*
>
> *Yours sincerely*
>
> *Bodil Holst*

Both of these papers were accepted and the *Journal of Microscopy* paper did indeed make it to a cover image [35]. The cover letter to the *Journal of Microscopy* is very short, the paragraph describing what the paper is about is just two sentences. This is because it is a very clear result which is obviously relevant for the journal, so no need for more than that.

The cover letter to *Scientific Reports* is longer, perhaps a bit too long. (It did fit on one page though.) I have included it here because I think it manages quite nicely to present the paper and the authors in the most positive light possible. This was particularly important here because I was worried the paper might be considered out of scope for the journal, so I made a big point of mentioning that that they had already published a similar paper from us and that this paper had actually been quite successful. The very subtle hidden message in the last paragraph is of course that this paper will get you more than 6000 views too if you accept it. I did not promise too much, I am happy to say. The paper was accepted and has now reached more than 12000 views [50].

12.2. Making it Past the Editor's Wastepaper Basket

The top journals (*Nature*, *Science*, etc.) only publish about 10% of the manuscripts submitted. Most submitted manuscripts are not sent to external referees, but are filtered out by a professional editorial staff. It is my impression that these people are very conscientious and do an excellent job. Nonetheless, you may have the bad luck of running into a novice editor having a bad day and too much work to do. On that particular day, some of the manuscripts on that particular editor's desk may land a little too quickly in the wastepaper basket. This is where being able to write in your cover letter that you are a Nobel Prize recipient comes in immensely handy. On balance, one may assume that Nobel Prize recipients have a better chance of making it past the editor's wastepaper basket than non-Nobel Prize recipients! Being from Harvard, MIT or Cambridge probably helps too, although it is by no means a guarantee. So if you do not have the initial psychological advantage of being a Nobel Prize recipient or belonging to a famous university/institution, do your best to convince the editor that your work is important by writing in the cover letter that you are submitting this paper on the recommendation of somebody famous.

- I was inspired to the text above by a friend of mine. While working as a postdoc he got a clever idea, somewhat outside his normal field. He pursued the idea and wrote up the paper as a single author. To make sure that everything was OK he approached a famous expert and asked his opinion on the paper. The famous expert was thrilled and recommended my friend to publish his paper in a very good journal (impact factor 20 or so). The paper was rejected within a week. One can easily imagine what happened: The editor saw a single author paper from a person not known in the field, working at an institute also not known in the field - Not a good starting point. My friend went back to the famous expert, who said, *"Rejection, oh that's strange; did you write in the cover letter that I recommend it for publication?"* *"No"*, was the reply (and the thought was, *"a bit late to suggest that now"*). *"Oh well, never mind."*, the conversation continued, *"Submit it to journal X instead. I am an editor there and I will see to it that it gets through."*. Journal X had impact factor 15 or so; the paper was duly submitted and very quickly accepted.

- Needless to say you should only include the recommendation from an expert in the cover letter if you have his or her explicit approval. Also, for it to make sense, the expert must really be an internationally renowned expert.

- The famous expert that my friend contacted had the reputation of being a fair and honest person. In general you must be very cautious when you approach somebody you do not know and show them your unpublished work. Do not assume they are nice, generous guys (or gals) just because they are famous. Some people have promoted their scientific careers largely by not being nice.

- It is not always possible to get a recommendation from a famous person. Of course you should submit your paper to the high-ranking journal anyway.

Part IV.
Publishing the Paper

Chapter 13.

Possible "Return States" of a Submitted Paper

The first step towards getting a paper published is of course to submit it as described in the previous part of the book. When you have submitted the paper, you will usually get an automatically generated email from the journal within a very short time (about 24 hours) confirming that they have received your manuscript. At the same time an email will be sent out to your co-authors asking them to protest if they have been included as co-authors against their will. If you do not receive a confirmation email within a week, check your spam-filter and if this does not yield anything, contact the journal and make sure everything is OK.

The refereeing process starts after the first submission confirmation and eventually you will get a response from the journal. If you have not heard anything more within two months after the first confirmation email, then it is perfectly acceptable to contact the journal and ask them how the reviewing of your paper is progressing. In many cases the journal submission system allows you to log into an author database and see how the work on your paper is progressing.

There will usually be 7 different possible initial "return states" for your paper. For a discussion on how to handle the resubmission of a paper, see Chapter 14.

- Your paper is rejected essentially immediately (usually within a week) by the editorial staff without being sent to external refereeing.

  – This is the typical situation arising when you try for a really high-ranking journal like *Nature* or *Science* or others. You will receive a standard reply, telling you something along the lines of: *"We do not want to publish your paper, but you should not see this as a criticism of your work, which we are sure is great; it is just not interesting enough for us to publish."*

Chapter 13. Possible "Return States" of a Submitted Paper

- Getting this kind of rejection is unpleasant, but normal. The good thing about the high-ranking journals is that at least waiting for the reply does not cost you much time .

- The editors of journals like *Nature* and *Science* and are not perfect nor do they claim to be. If you are really convinced of your work, then it is perfectly possible to resubmit your paper again after this initial rejection with an extended cover letter explaining that you really think this is something they ought to look into in detail. I did this myself with the paper "*Poisson's Spot with Matter Waves*" [36]. *Nature* still rejected it, but we got a much more personal rejection letter the second time round, in which they asked us to contact them when the paper was published elsewhere, because they would then consider including it in the *Research Highlights Section*, which they eventually did [37]. So in the end it was well worth it, and my former PhD student Thomas Reisinger, the driving force behind the work, got some well-deserved credit at a crucial stage in his career.

- Your paper is accepted straight away, you do not have to do anything (happens very rarely, but cases have been known).

    - Everything is fine. Count yourself lucky.

- Your paper is accepted subject to "*minor revisions*" or "*mandatory revisions*".

    - Everything is fine - "*minor revisions*" usually means things like typos, missing references, suggestions for slight changes to figures, etc.. "*Minor revisions*" implies something which you can easily fix. "*Mandatory revisions*" is used here in the sense that the revisions needing to be done are more severe. They do not necessarily require more work than the "*minor revisions*" but they are necessary to ensure that the paper is OK. This could, for example, relate to incorrect use of certain vocabulary or also calculation errors.

    - The two words "*minor*" and "*mandatory*" are not used in a very consistent manner. Do not worry about them; just focus on what you are actually told to do.

    - This situation is really the normal "best case". Even if they think the paper is nearly perfect, most referees will still suggest minor corrections, if for no other reason than demonstrating to the editor that they have really read the paper.

- * I have a friend, whose name shall not be mentioned here, who claims that he deliberately includes minor mistakes in his papers. His argument is that if he makes it easy for the referee to find a few small mistakes that they can ask him to correct, then they will not start looking more closely and perhaps discover some big mistakes. I pointed out to him that I would want the referees to find the big mistakes and that I would not want a paper with big mistakes in it published, but this was not his line of thought.

- The paper is accepted subject to "*major revisions*".
  - This actually happens quite rarely in my experience. If there are "*major revisions*" to make, the editors will usually not promise straight away to publish the paper. "*Major revisions*" can include nasty requests to do more experiments or calculations.

- You are asked to reply to the referee comments and invited explicitly to resubmit upon revision of your manuscript.
  - This is what usually happens in the case of "*major revisions*". In such cases your paper will usually be accepted in the end if you do a good job of addressing the referee comments (see Chapter 14). Sometimes it can bounce back between you and the referees several times, but you will get there in the end.

- Your paper is rejected, but you are told that if you can refute the referee comments, the editor might reconsider.
  - Don't give up! **Do not take no for an answer! Your interest in getting the paper accepted is greater than the referees' interest in rejecting it. Fight back and "tire them out"!**
  - Take a careful look at the referee comments.
    * For high-ranking journals, they will typically say that the results are only of interest to a very specialized, scientific audience. Then you rewrite your introduction with more and broader references and in particular recent references in high-ranking journals.
    * If the objections are more technical and you have some good replies, try to resubmit and include them.

Chapter 13. Possible "Return States" of a Submitted Paper

> > * Generally, the more the referees have written, the better your chances are! Lengthy referee comments are generally a good thing even if the tone is negative. It shows that the referees have really taken an interest in your work. Surprisingly, this type of referee can often be "won over" by addressing their comments carefully (see Chapter 14).
>
> – Sometimes you will have one good and one bad referee. In this case, address all the referee comments, resubmit and in addition write in the reply letter (see Chapter 14) that if the editor still does not want to accept the paper, you request a third referee before a final decision is taken.
>
> – If you think one of the referee comments does not live up to the scientific standard, you still revise the manuscript addressing all comments as well as you can, but you also do complain directly to the editor and request that you are given another referee unless the editor is willing to accept the paper immediately.

- Your paper is rejected flat out as not suitable for publication in this journal.

  – Also here the general rule is that you should not despair, but study the comments of the referees carefully and address them as described above.

  > * There can be situations however, where there is no hope and no point in wasting your or the editors' time. As an example of this, I'd like to mention how our *Journal of Microscopy* cover story, *Nature* research highlight paper on helium microscopy (see Chapter 12), which we initially tried to publish in *Physical Review Letters*, was rejected by the journal. The referee comments essentially all said the same, namely that the paper did not contain any new physics, since focusing of atoms had already been demonstrated in previous publications, and the mere obtaining of a poor resolution image was an "*obvious extension*". This "*obvious extension*" had caused us months and months of hard work, but the referees were in a sense right and I realized that I had been trying to publish in the wrong journal.

Chapter 14.

Resubmission or How to Address the Referee Comments?

In this chapter I discuss how to handle the resubmission of a paper upon having received the referee comments. As discussed in the previous chapter, there are situations in which it is not possible to resubmit a revised manuscript to the same journal. In that case, you have to find a new journal and submit the manuscript again. Even so, I recommend that you address the referee comments before you submit the manuscript to the new journal. If for no other reasons than because you might get the same referee(s) again. (I have heard of surprisingly many cases of this happening.) **If you resubmit your paper to a new journal after it has been rejected by another journal, do NOT tell the new journal that you have previously tried to publish your manuscript elsewhere.** This is basic psychology, nobody is very keen on accepting what other people have rejected.

The refereeing process can be both painful and frustrating and it is not always completed to satisfaction, in the sense that a paper does not always get published in the journal where you originally wanted it to be. However, judging from my own experience, I can say that I have not had a single paper which did not improve through the refereeing process. This includes the ones I had to resubmit to new journals.

It is important to realize that **everybody gets bad referee comments at some point, sometimes VERY bad comments.** It seems that some colleagues use the refereeing process as an outlet for their general frustration and dissatisfaction with life, the universe and everything. It is important to remember that **this is NOT (necessarily) something which reflects on the quality of your work**. Quite recently a paper of mine was initially rejected, perhaps because one referee wrote: "*There is nothing new or enlightening in this article.*". We fought back! The topic of the paper was how you can use a low energy electron beam lithography device for fast, large area writing. The main complaint of the referee was that large area writing can be done much better with high energy electron beam lithography. This is true,

Chapter 14. Resubmission or How to Address the Referee Comments?

but high energy electron beam lithography devices are much more expensive. After a bit of discussion with the editor, whereby we, among others, pointed out that the referee appeared rather biased towards high energy electron beam lithography, the paper ended up being published in that same journal with very minor revisions [51]. OK, we had to resubmit it as a "shop note" (see Section 3.1) rather than a full paper, but who cares! Actually a shop note was more appropriate. For several months it was one of the most downloaded papers from the journal website.

ADDRESSING REVIEWER COMMENTS — BAD REVIEWS ON YOUR PAPER? FOLLOW THESE GUIDELINES AND YOU MAY YET GET IT PAST THE EDITOR:

Reviewer comment:
"The method/device/paradigm the authors propose is clearly wrong."

How NOT to respond:
✗ "Yes, we know. We thought we could still get a paper out of it. Sorry."

Correct response:
✓ "The reviewer raises an interesting concern. However, as the focus of this work is exploratory and not performance-based, validation was not found to be of critical importance to the contribution of the paper."

Reviewer comment:
"The authors fail to reference the work of Smith et al., who solved the same problem 20 years ago."

How NOT to respond:
✗ "Huh. We didn't think anybody had read that. Actually, their solution is better than ours."

Correct response:
✓ "The reviewer raises an interesting concern. However, our work is based on completely different first principles (we use different variable names), and has a much more attractive graphical user interface.

Reviewer comment:
"This paper is poorly written and scientifically unsound. I do not recommend it for publication."

How NOT to respond:
✗ "You #&@*% reviewer! I know who you are! I'm gonna get you when it's my turn to review!"

Correct response:
✓ "The reviewer raises an interesting concern. However, we feel the reviewer did not fully comprehend the scope of the work, and misjudged the results based on incorrect assumptions."

JORGE CHAM © 2005

www.phdcomics.com

14.1. Some Real Examples of Referee Comments.

I am grateful to my young, successful colleague, who very kindly allowed me to show you the referee comments below. They are all real!

A good day

- *"I would like to congratulate the authors.."*

- *"This is a very well-written manuscript on an important problem in current research."*

- *"The paper is very readable. It is interesting and should be understandable by a wide readership."*

- *"It also has the potential to stimulate additional experimental work."*
- *"I therefore recommend the publication of this article in Physical Review Letters as it is."*

A bad day

- *"The authors clearly demonstrate their complete lack of understanding of the problem."*
- *"The value of this manuscript is zero."*
- *"The manuscript should not be published in this or any other journal."*

14.2. The Reply Letter

When you resubmit your manuscript to a journal, you have to provide a revised manuscript together with a reply letter to the editor, explaining how you have addressed the referees comment. There are no set rules for how this letter should be structured, the main thing is that **it must be easy for the editor to understand how you have addressed the referee comments in your revised manuscript. Further, and very importantly: ALL issues raised by the referee(s) must be addressed in the letter. Do not ignore one single point, no matter how stupid or annoying! Ignoring referee comments may be seen as disrespectful or taken as a sign of incompetence. You do not want either of that.** In Section 14.2.2 you will find a list of typical, annoying referee comments and suggestions for how to handle them. **Equally important: Do not address any issues NOT pointed out by the referee(s) unless you spot a serious error or trivial printing/spelling mistakes.** The reason for this is perhaps not quite so obvious, but the point is that if you make changes not related to the referee comments then the paper differs from the paper the referees evaluated, and which paper is it then that should be accepted for publication? and by whom? Once a paper is submitted you must consider it as finished and only do what you are told by the referee(s) and editor.

I present here the structure I use for the reply letter. An example of one of my reply letters can be found in Section 14.2.1.

I always split the reply letters into two parts:

Chapter 14. Resubmission or How to Address the Referee Comments?

- The first part of the reply letter is addressed directly to the editor handling your paper and contains the following:
  - A header with the identification number of your paper (assigned by the journal) and the title of the paper.
  - A paragraph where you express your thanks for receiving the referee comments and highlight the most positive replies from the referees (hopefully there are many!). The strategy behind this is to remind the editor that this is really a paper which the referees think is great and should be published.
  - A paragraph with comments that you do not wish the referee(s) to see, for example if you find that one of the referees has been unjust. Technical errors should not be addressed here, unless you find that the referee(s) have done such a bad job that they seem to be completely incompetent, in which case you can write a sentence like: *"As can be seen in our reply to the referees below, referee 2 demonstrates a severe lack of knowledge on the subject matter treated in this paper"*.

- The second part of the reply letter is headed: *"Reply to the referees"*.
  - Whatever you do when you write your reply addressed directly to the referee(s), be polite (even if they have misunderstood everything). Referees usually work without payment and editors often have a hard time finding suitable referees, so they do not want rude authors to upset them. A good way of starting the *"reply to the referees"* section is: *"We thank the referees for their (insightful/useful/helpful) comments"* - words in parenthesis can be omitted if you are really irritated with the referees.
  - Every issue pointed out by the referee(s) must be addressed, even if they are clearly wrong on a particular issue you cannot simply ignore them. See Section 14.2.2 for some suggestions on how to handle typical, nasty referee comments.
  - I find that a good way of writing the *"reply to the referees"* is simply to take each referee comment and split it up in sections, write these sections in bold font and provide my comments to each section in between in non-bold font. This makes a very clear layout for referee(s) and editor to follow.
  - At the very end you can include a section *"Additional Changes"*. Here you list changes you have done NOT related to the referee comments. Remember, only trivial spelling mistakes or serious errors can be addressed at

this stage (see discussion above). You do not have to list all the places in the text where you have corrected spelling mistakes, a short comment "*We found and corrected a few additional spelling mistakes*" suffices. Serious errors should be discussed in more detail.

14.2.1. An Example of a Set of Real Referee Comments with Reply Letter

Below you can read the referee comments and reply letter for one of the papers that I have co-authored. I have chosen this correspondence, reproduced here in its full length, because the topic of the paper allows the referee comments to be more easily followed by a general scientific audience than is the case for most of my other publications. This correspondence is very typical, including the request for more work and the fact that one of the referees has actually misunderstood something quite essential.

Referee Comments

Manuscript ID ARCH-04-0071-2013 entitled "Flax look-alikes: Pitfalls of ancient plant fiber identification" which you submitted to Archaeometry, has been reviewed. The comments of the referee(s) are included at the bottom of this letter. The referee(s) have recommended major revisions to your manuscript. Therefore, I invite you to respond to the referee(s) comments and revise your manuscript.

Comments to the author: Both reviewers have said that this is an important contribution, but reviewer #2 has asked for the paper to be reorganized, and, if, possible, an example given. I can see that this might be difficult, but please consider this suggestion.

Referee: 1 For the textile archeology community this is a highly significant and salutary, critical, review of the standard physical methods of analyzing and identifying the principal bast fibers. It is argued here, and apparent, that NONE is reliable. The earliest history of use of bast fibers for rope, textiles etc is a controversial issue. This paper and its figures demonstrates that the analytical methodology usually applied will not stand up to scrutiny. A breakthrough into new approaches is urgently required. There are a few typographical errors. The bibliography seems to be incomplete.

Chapter 14. Resubmission or How to Address the Referee Comments?

Referee: 2 This brief paper presents useful observations on the differentiation of plant fibers from hemp, flax and nettle, and is potentially of wide interest for the study of ancient textiles. The structure of the paper needs to be revised, however, and it would also benefit from inclusion of a working example of archaeological application and/or re-interpretation.

Specific comments: -3. Results - the first sentence of the section interprets the results before they have been presented. The authors should first present their observations and then infer their implications subsequently. - Section 3.1 - 'Ref Bergfjord' needs to be filled in -3.4 Flax look-alikes - since these are potentially 'outliers' or 'one-off' examples of flax-like fibers it is important to explore the implications of the study further by (re-)applying the new criteria to a published assemblage, in order to illustrate how the new method works in practice and how it can re-direct interpretation of ancient plant fiber use.

Reply Letter

To Professor X, Editor, Archaeometry

Concerning: Resubmission of Manuscript "Flax look-alikes: Pitfalls of ancient plant fiber identification", Manuscript ID ARCH-04-0071-2013 Authors: E. Haugan and B. Holst

Dear Professor X

Thank you very much for your email 10 days ago containing the referee comments to our paper. We are delighted that both referees acknowledge this as an important contribution. Below you will find a detailed overview of how we have addressed the referee comments and revised the manuscript. We have addressed all issues commented on by the referees.

Reply to the Referees:

We thank the referees for their insightful comments. For clarity the referee comments are included as sections below (written in bold) with our comments to each issue listed in between.

Referee 1:

For the textile archeology community this is a highly significant and salutary, critical, review of the standard physical methods of analyzing and identifying the principal bast fibers. It is argued

here, and apparent, that NONE is reliable. The earliest history of use of bast fibers for rope, textiles etc is a controversial issue. This paper and its figures demonstrates that the analytical methodology usually applied will not stand up to scrutiny. A breakthrough into new approaches is urgently required.

We thank Referee 1 for these very positive comments, which acknowledge the importance and relevance of our manuscript.

There are a few typographical errors.

We have scrutinized the manuscript for typographical errors and corrected them.

The bibliography seems to be incomplete.

We are a bit puzzled about this remark. The literature on ancient plant fiber textiles is vast and it would clearly be beyond the scope of this paper to include all references. We have conducted a new literature search as a response to this comment, but we remain confident that we have included the most important references relevant for this paper, both in terms of primary literature on characteristic traits for plant fibers as well as on new and established methods for reliable plant fiber identification. If we have missed anything important, we would be grateful for the reviewer to point this out to us specifically.

<u>Referee 2:</u>

This brief paper presents useful observations on the differentiation of plant fibers from hemp, flax and nettle, and is potentially of wide interest for the study of ancient textiles.

We thank Referee 2 for these very positive comments, which acknowledge the importance and relevance of our manuscript.

The structure of the paper needs to be revised, however

We have revised the structure according to the specific comments below.

and it would also benefit from inclusion of a working example of archaeological application and/or re-interpretation.

While we appreciate the referee's interest in the investigation of archaeological samples, this would be beyond the scope of this paper and remove the focus from the main point: methodology relevant for Archaeology (hence the submission to Archaeometry). Further, we are not archaeologists and hence it is hardly appropriate that we embark on deciding on which archaeological samples it may be relevant to analyze, let alone re-interpret. We simple cherish the hope that this paper will inspire archaeologists to re-investigate what they decide are important samples.

Specific comments: -3. Results - the first sentence of the section interprets the results before they have been presented. The authors should first present their observations and then infer their implications subsequently.

We have removed the first sentence in section 3 (Results).

-Section 3.1 - 'Ref Bergfjord' needs to be filled in

This has been done. We apologize.

-3.4 Flax look-alikes - since these are potentially 'outliers' or 'one-off' examples of flax-like fibers

Actually, as pointed out in section 2, non-typical fibers were quite common in the fiber material that we investigated. It is possible that the amount of non-typical fibers may vary with growth conditions etc., but since the growth conditions are unknown anyway when archaeological samples are investigated, this is not relevant here.

it is important to explore the implications of the study further by (re-)applying the new criteria to a published assemblage, in order to illustrate how the new method works in practice and how it can re-direct interpretation of ancient plant fiber use

We are a bit puzzled about these remarks. We neither propose "new criteria" in this manuscript nor do we introduce any "new method". We simply show that the criteria commonly used may not lead to a correct identification. Hence we do not see what measurements we can reasonably present in this paper beyond those already presented.

Yours sincerely

Bodil Holst

The paper was accepted without further revisions. The "major" revisions suggested by the editor turned out to be very minor.

14.2.2. How to Handle Some Typical Referee Comments

Of course the referee comments will vary a lot from field to field, but there are some typical referee comments which apply to all fields. Here they are, with suggestions on how to handle them.

- The referee wants you to do more work, i.e., more experiments, more data analysis, more calculations or similar - nasty!
  - Don't do more experiments or data analysis or calculations unless it is very easy, which it usually isn't or you would probably have done them already. Argue as well as you can why what you have is good enough.
    * Anything which takes you more than two weeks is not "very easy".
    * It is OK to refer to experimental reality when you argue against more experiments: temperature drift in the laboratory, the instrument has been rebuilt since taking the data, etc.. It is also OK to be a little bit creative with reality and make certain reasons appear worse than they perhaps really are. The main thing is to come up with something with which to address the referee comment without actually doing what he/she asks.
      · I have the sneaky suspicion that the "*do more experiments*" comment is particularly typical when theorists are asked to review experimental papers.
    * Unless very convincingly argued by the referee, too little data will not normally be a reason for rejection by the editor.
    * NB! There might be the rare case where the referee really points out something essential that you have overseen. In that case you just have to bite the bullet and do the work regardless of how long it takes.
  - Do not do what one of my workshop participants did. After having received the comment "*do more experiments*" she spent another 6 months in the lab before she resubmitted the paper. I asked her if this further work had changed any of the scientific conclusions of the paper and she said it had not. Fortunately she could still publish the paper. It could easily have happened that somebody else might have published a paper on the same topic within those 6 months, and then it would really have been a tragedy.
  - Sometimes you can use "*beyond the scope of the paper*" as an excuse for not doing some suggested work, see the reply letter in Section 14.2.1.

Chapter 14. Resubmission or How to Address the Referee Comments?

- The referee has misunderstood something - tricky!
  - Remember that you must always be polite, but of course you should not change something which is correct into something which is wrong. A possible response might look like this:
    * The referee points out that our conclusion on page X is incorrect.
    * Your response: *"We regret that we have not been able to express ourselves in a clear manner and have reformulated the conclusion accordingly"*.
    * If the misunderstanding is really severe, you can be more direct as in the reply letter in Section 14.2.1, but please **NEVER write:** *"the referee has misunderstood something"*.

- The referee wants you to include some specific, additional citations - easy!
  - Include the citations that the referee suggests, even if you do not think they are highly relevant. Do not argue this point! It is not uncommon that referees ask to have their own papers cited as a small reward for their refereeing work.

- The referee complains that your English is not good even though it is perfectly fine - irritating!
  - This actually happens quite often, so let it not disturb you too much. It even happened to me once with a paper written together with a native English speaker with an excellent style of writing. I recently submitted another paper where one referee complained that *"in some sections the general readability of the paper is low"* whereas the other referee wrote that *"the paper is well written and the presentation and general discussion of the results is very clear"*!
  - I suspect that "grumpy" comments about the English are often due to the referee(s) having problems understanding what you have written. Sometimes the message you try to convey is complex and in particular if the referee(s) are not completely up to date on the field it might be difficult for them to follow, even if the text is well written. However, they do not want to admit this to themselves and so the writing must take the blame. In the last example above it was clear from the rest of the comments that

14.2. The Reply Letter

the second referee had a much better understanding of the paper than the first referee.

– So what to do?

* A diplomatic response is called for. Obviously you should not rewrite a well-written paper. You simply read through it and modify the text a bit. You have not seen the text for several weeks and actually a very good way to improve a text is to leave it completely for a few weeks and then take it up again. So take this as a welcome opportunity that you would not otherwise have had to improve the style of your paper. You do not have to list every single change you have made in the reply letter, just write a comment like: "*We have read through the manuscript carefully and modified the language in several places*".

* Double check that you are using correct spelling and the spelling that the journal prefers. If the referee(s) complain about spelling mistakes it can be because they are expecting American spelling rather than English or vise versa. If you have used the wrong spelling for the journal, correct it and write that you have changed from English to American spelling or vise versa. If you have not made a general spelling mistake simply write: "*We have checked and corrected the manuscript for spelling mistakes*". This sentence you can use even if you do not find a single spelling mistake.

- The referee wants you to write an "old-style" abstract (see Section 5.3) - do not give in!

– Some "old-fashioned" referees are not so keen on the modern "free-style" abstract which includes introductory remarks to the topic at the very beginning: "Why is this important? (see section 5.3).

– Usually my advice is "*give the referees what they want, just get the paper published*", but here you should not give in, because it may mean that potential readers coming from other fields than your own miss your paper as relevant for their work. Do not change anything! Formulate a reply along the following lines:

* *The referee complains that the introduction to our abstract is too long. We point out that our abstract does not exceed the allowed length. Further, we find that a couple of introductory remarks are appropriate in*

Chapter 14. Resubmission or How to Address the Referee Comments?

> *the abstract to generate interest in a larger, potential readership including readers that do not have online access to the journal and need to estimate the relevance of the paper on the basis of the abstract alone.*
>
> · Of course the last part of the last sentence does not make sense if the journal is open access, then you just stop after *"potential readership"*.

- You have included a paper-outline (see Section 5.5.3) at the end of the introduction and the referee does not like it - go ahead and give in!

  – As mentioned in Section 5.5.3 there appears to be two schools here. Some people like the paper outline at the end of the introduction and some people don't. If you happen to come across a referee who does not like it and complains, just take it out. It is not so important.

- You have two referees who contradict each other - confusing!

  – For example one referee wants you to include an additional figure, another complains about there being too many figures. Above you saw the example where one referee says the English is good while the other concludes that it needs changing.

  – So what to do?

    * Do not worry too much, usually this is not such a big deal. Especially because if your paper is sent back to the referees for a second round, they will get the chance to see the comments of each other and see for themselves that there is a contradiction. Still diplomacy is called for in the reply letter. You must try, nonetheless, to satisfy both referees. In the case of the figures, you can remove one figure and add the information requested by one referee to an already existing figure, for example. If there is no other way of doing it, make two figures, label them "a" and "b" and put them in one figure. Then you have made one more figure without making one more figure. In the case of English, modify the text a bit as described above; the other referee is not going to complain about that.

- On a more general level one referee may recommend the paper for publication while the other does not.
    * In this case you emphasize the positive referee comments in the first part of the reply letter and request as a "minimum" option that the paper is sent to a third referee if the editor is not willing to publish it straight away: "*We note that referee 1 has very positive comments and recommends our paper to be published in this journal. We have addressed all comments from both referees and trust that referee 2 will now also be satisfied. Should this not be the case, we request that the paper is sent to a third referee.*"

Chapter 14. Resubmission or How to Address the Referee Comments?

14.3. Being a Referee Yourself

A very good way to gain understanding of the refereeing process is to be a referee yourself. Being a referee is also important for your CV, because it serves as an important quality stamp that you have been accepted in your research community. The more prestigious the journals you can list, the better.

It is quite common that professors ask their students or postdocs to assist in the refereeing process. (This is not something you can put on your CV.) If you are a student or a postdoc it is fine to accept this task a couple of times, especially if you get to see the final version your professor submits and get some helpful comments from him/her. However, if this is not the case or if you are asked to review more than 3 papers, then tell him/her in a friendly manner that you know it is possible also for PhD students and postdocs to act officially as referees and you would very much appreciate it if he/she would be so kind as to suggest you as an official referee for the journal. With a bit of luck he/she will do just that or at least (hopefully) not bother you with more unacknowledged refereeing tasks.

Part V.
Final Remarks

Chapter 15.
A Small, Final Reflection

This book is now coming to an end. I hope you have enjoyed reading all or part of it. I have tried to provide you with two things: 1) a practical tool kit for writing papers yourself and 2) an honest insight into the world of scientific publishing. I hope I have succeeded in this. More importantly, I hope I have succeeded in encouraging you in your paper writing work. Please remember that even though paper writing can be difficult and the whole publishing process sometimes painful and frustrating, your published papers are your gift to mankind, your contribution to bringing this world forward. Your papers will remain when you have gone (as long as you have published in a proper, established journal of course). This is a privilege and a responsibility!

I wish you good luck and happy writing!

Bibliography

[1] Binnig, G. & Rohrer, H. *Nobel Lectures in Physics, 1981-1990*, 383–412 (World Scientific Publishing Company, Singapore, 1993).

[2] Güntherodt, G. Nobel prize in physics 2007: Peter Grünberg. *Europhysics News* **39**, 8 (2008).

[3] Singer, C., Holmyard, E., Hall, A. & Williams, T. (eds.) *A History of Technology*, vol. 3 (Oxford University Press, Oxford, 1958).

[4] http://royalsociety.org/about-us/history/.

[5] http://royalsociety.org/about-us/governance/president/.

[6] Chapman, A. England's Leonardo: Robert Hooke (1635-1703) and the art of experiment in Restoration England. *Proceedings of the Royal Institution of Great Britain* **67**, 239–275 (1996).

[7] Hooke, R. *Micrographia: or some Physiological Descriptions of Minute Bodies made by Magnifying Glasses* (J. Martyn and J. Alestry, London, 1665).

[8] Oldenburg, H. The introduction. *Philosophical Transactions* **1**, 1–2 (1665).

[9] Newton, I. *Philosophiae Naturalis Principia Mathematica* (The Royal Society of London for Improving Natural Knowledge, 1687).

[10] Ørsted, H. C. Experimenta circa effectum conflictus electrici in acum magneticam. *Journal für Chemie und Physik* **29**, 275–281 (1820).

[11] Whitesides, G. M. Whitesides' group: Writing a paper. *Advanced Materials* **16**, 1375–1377 (2004).

[12] Garfield, E. The history and meaning of the journal impact factor. *The Journal of the American Medical Association* **295**, 90–93 (2006).

[13] http://thomsonreuters.com/products-services/science/free/essays/journal-selection-process/.

[14] West, J., Bergstrom, T. C. & Bergstrom, C. The eigenfactor metrics: A network approach to assessing scholarly journals. *College and Research Libraries* **71**, 236–244 (2010).

[15] Hirsch, J. E. An index to quantify an individual's scientific research output. *Proceedings of the National Academy of Sciences* **102**, 16569–16572 (2005).

[16] Poole, A. The independent obituary: Jeremy Maule (2011). URL http://www.independent.co.uk/arts-entertainment/obituary-jeremy-maule-1188609.html.

[17] Oreskes, N. & Conway, E. M. *Merchants of Doubt* (Bloomsbury Press, New York, 2010).

[18] Bergfjord, C. *et al.* Nettle as a distinct bronze age textile plant. *Scientific Reports* **2**, 664 (2012).

Bibliography

[19] Hanson, B. Science editors choice: Wild textiles. *Science* **338**, 305 (2012).

[20] Bohannon, J. Who's afraid of peer review? *Science* **342**, 60–65 (2013).

[21] http://bogus-conferences.blogspot.no/.

[22] http://www.nlm.nih.gov/pubs/factsheets/medline.html.

[23] Lin, T. Cracking open the scientific process. *The New York Times* **Jan. 17**, D1 (2012).

[24] Eder, S. D. *et al.* Focusing of a neutral helium beam with a photon-sieve structure. *Physical Review A* **91**, 43608 (2015).

[25] Gustavii, B. *How to Write and Illustrate a Scientific Paper* (Cambridge University Press, Cambridge, 2003).

[26] Pitnick, S., Spicer, G. S. & Markow, T. A. How long is a giant sperm? *Nature* **375**, 109 (1995).

[27] Goldacre, B. Will asking a question get your science paper cited more? (2011). URL http://www.theguardian.com/commentisfree/2011/oct/14/does-a-question-get-science-paper-cited.

[28] Stankovich, S. *et al.* Graphene-based composite materials. *Nature* **442**, 282–286 (2006).

[29] Carlson, C. M. *et al.* Kruppel-like factor 2 regulates thymocyte and T-cell migration. *Nature* **442**, 299–302 (2006).

[30] Alleman, M. *et al.* An RNA-dependent RNA polymerase is required for paramutation in maize. *Nature* **442**, 295–298 (2006).

[31] Xing, H., Takhar, P. S., Helms, G. & He, B. NMR imaging of continuous and intermittent drying of pasta. *Journal of Food Engineering* **78**, 61–68 (2007).

[32] Bilgicli, N., Ibanoglu, S. & Herken, E. N. Effect of dietary fibre addition on the selected nutritional properties of cookies. *Journal of Food Engineering* **78**, 86–89 (2007).

[33] Sutar, P. P. & Gupta, D. K. Mathematical modeling of mass transfer in osmotic dehydration of onion slices. *Journal of Food Engineering* **78**, 90–97 (2007).

[34] Steurer, W. *et al.* Observation of the boson peak at the surface of vitreous silica. *Physical Review Letters* **99**, 035503 (2007).

[35] Koch, M. *et al.* Imaging with neutral atoms: a new matter-wave microscope. *Journal of Microscopy* **229**, 1–5 (2008).

[36] Reisinger, T. *et al.* Poisson's spot with molecules. *Physical Review A* **79**, 53823 (2009).

[37] Research highlights: Seeing spots. *Nature* **459**, 486 (2009).

[38] Shifman, M. ITEP lectures in particle physics. *arXiv:hep-ph/9510397* (1995).

[39] Ward, B. F. L. Relativ size of SU(3) symmetry-breaking effects and penguin diagrams. *Physical Review Letters* **70**, 2533–2536 (1993).

[40] Hyldgard, A. *et al.* Fish & chips: Four electrode conductivity/salinity sensor on a silicon mulit-sensor chip for fisheries research. *Sensors IEEE* 1124–1127 (2005).

Bibliography

[41] Aad, G. *et al.* The ATLAS experiment at the CERN Large Hadron Collider. *Journal of Instrumentation* **3**, S08003 (2008).

[42] Gregory, S. G. *et al.* The DNA sequence and biological annotation of human chromosome 1. *Nature* **441**, 315–21 (2006).

[43] Bergfjord, C. *et al.* Comment on "30.000-year-old" wild flax fibers. *Science* **328**, 1634–b (2010).

[44] Kvavadze, E. *et al.* 30.000-year-old wild flax fibers. *Science* **325**, 1359 (2009).

[45] Wennerås, C. & Wold, A. Nepotism and sexism in peer review. *Nature* **387**, 341–343 (1997).

[46] Watson, D., Andersen, A. C. & Hjort, J. Mysterious disappearance of female investigators. *Nature* **436**, 174 (2005).

[47] Reich, E. S. *Plastic Fantastic: How the Biggest Fraud in Physics Shook the Scientific World* (Palgrave MacMillian, New York, 2010).

[48] http://dx.doi.org/10.1016/j.amc.2007.03.011.

[49] Van Noorden, R. Publishers withdraw more than 120 gibberish papers. *Nature News* (2014).

[50] Skoglund, G., Nockert, M. & Holst, B. Viking and early middle ages northern scandinavian textiles proven to be made with hemp. *Scientific Reports* **3**, 2686 (2013).

[51] Greve, M. M. & Holst, B. Optimization of an electron beam lithography instrument for fast, large area writing at 10 kV acceleration voltage. *Journal of Vacuum Science & Technology B* **31**, 043202 (2013).

Index

Abbreviations, 54
Abstract, 47, 53–55, 69, 82, 83, 95, 97, 112, 124, 131, 141, 167
Abstract, conference, 28, 30, 136
Abstract, examples, 74
Acknowledgments, 52, 53, 62, 86, 100, 101, 113, 132
Acronyms, 54
Appendices, 52, 53, 105
Article, *see* Paper
Author, 3, 5, 7, 8, 16, 22, 27, 32, 33, 53, 61, 84, 87, 88, 100, 115, 122, 142, 145, 148, 149, 153, 160
Author contribution statement, 101
Author ID code, 68
Author Name, 67
Author, co-author, 16, 43, 61, 63, 100, 107, 112, 132, 134, 143, 153
Author, corresponding, 66, 84, 145
Author, first, 11, 59, 62–65, 101
Author, last, 63

Citation, 23, 29, 30, 32, 47, 62, 70, 76, 81, 82, 84, 102, 103, 112, 132, 144, 146, 166
Citation, self-citation, 14, 16, 19, 61, 102
Conclusion section, 47, 52, 53, 76, 96, 97, 112

Conference proceedings, 13, 21, 23, 28, 29, 126
Conference Talk, dos and don'ts, 136
Conference Talk, objections to give, 135
Consistency, 76, 80, 93
Corrigendum, 22
Cover letter, 102, 142–145, 149, 150, 154
Cover letter, examples, 146

Database, *see* Search engine
Discussion, 53, 96
Discussion, importance of, 18, 28, 100, 109

Editorial board, 7, 27, 128, 143, 145
Elsevier, 5, 26, 128
English language, 6, 22, 114, 119, 122, 123, 125, 166, 168
English, sloppy, 113
English, worry about, 43, 112, 114, 135
Erratum, 22
European Commission, 27, 100
Experimental work, 41, 44, 65, 90, 93, 107

Figure caption, 95, 115, 116
Figures, 94, 95, 108, 111, 115, 133, 142, 168
Final version, approval of, 112

177

Index

Full paper, 20, 29, 141, 145, 158

German language, 125
Google Scholar, 16, 32

Habilitation candidate, 64–66
Hirsch index, 13, 15, 17, 31

Impact factor, 13, 15, 17, 18, 20, 23, 25, 27, 30, 32, 50, 102, 112, 128, 146, 150
Impact factor, how to calculate, 13
Interdisciplinary work, 31, 62, 80, 143
Introduction, 52, 53, 70, 80, 93, 103, 112, 120, 144, 155, 168

Journal, how to choose, 23

Key words, 79

Legal responsibility, your, 33, 102, 115
Letter, 20, 52, 71, 145

Master thesis, 66
MEDLINE, 32
Methods section, 53, 87, 112, 122

Nature, 5, 12, 15, 24, 25, 27, 30, 31, 34, 40, 56, 57, 59, 71, 73, 79, 103, 105, 107, 114, 122, 134, 149, 153
NLM, 32
Note paper, 21, 158
Number of papers, 12

Online supplementary material, 53, 90, 105
Open access publishing, 18, 26, 31, 32, 168
ORCID, 68

Paper format, 141
Paper structure, 22, 29, 52, 126, 128, 141
Papers, different types, 20
Patent process, 126
Peer reviewing, 6–8, 10–12, 40, 131, 134, 142, 143, 153, 157, 170
PhD student, 11, 21, 24, 44, 55, 59, 61, 63–66, 88, 107, 154, 170
PhD thesis, 11, 89, 107, 113, 116
PhDs in politics, 116
Postdoc, 11, 44, 63–66, 142, 149, 170
Previous work, 16, 70, 73, 81, 82, 93, 96
Previous work, dos and don'ts, 85
Professor, 3, 44, 45, 106, 110, 116, 170
Professor, assistant, 64–66
Publication, *see* Paper
PubMed, 32

Rapid Communication, 20, 145, 147
Referee, 24, 26, 62, 85, 87, 100, 105, 113, 120, 122, 133, 149
Referee comments, abstract style, 167
Referee comments, bad, 8, 157, 165
Referee comments, contradicting, 168
Referee comments, do more work, 165
Referee comments, English, 166
Referee comments, examples, 157, 158, 161
Referee comments, how to handle, 165
Referee comments, misunderstandings, 166
Referee comments, typical, 165
Referee, being one yourself, 170
Referees, are only human, 8, 29, 62, 68, 79, 82, 131, 132, 134

Referees, choosing, 142, 143, 146
Referees, get to know, 134
Reference, *see* Citation
Rejection, 24, 150, 154–156, 165
Reply letter, 159
Reply letter, example, 162
Researcher ID, 68
ResearchGate, 33
Resubmission, 24, 153, 157, 159, 162
Results, 6, 9, 11, 12, 17, 19–21, 29–31, 40, 41, 43, 44, 47, 49, 57, 69, 74, 75, 83, 85, 87, 93, 97, 107, 109, 111, 112, 117, 124, 136, 146, 155
Results and Analysis section, 52, 53, 122
Results, conflicting, 78
Results, negative, 50
Review paper, 21
Reviewer, *see* Referee
Revisions, major, 155, 161, 164
Revisions, minor, 154, 164

Science, 13, 27, 30, 31, 34, 59, 62, 103, 107, 114, 122, 148, 149, 153
Scientific board, 144, 146
Scientific prestige, 10, 12, 100
Search engine, 13, 32, 33, 54, 68, 79, 84
Short Communication, 20
Single and double blind reviewing, 7
Spelling, checking, 113, 159, 160, 167
State-of-the-art, 52, 70, 73, 81, 83, 102, 132
Submission process, 141
Summary paragraph, Nature's guidelines, 71
Supervisor, 11, 21, 42, 58, 63, 64, 107–109, 113, 136, 137

Take-home message, 43, 47, 49–51, 56, 57, 70, 73, 75, 78, 86, 95, 108, 111, 112, 121
Technical comment, 22, 62, 114
Theoretical background section, 53, 93
Time slots for writing, 41, 43, 108
Title, 30, 47, 53, 54, 78, 80, 82, 112, 120, 141, 145, 160

Web of Science, 13, 19, 22, 23, 30–32, 84

Printed in Great Britain
by Amazon